# DATE DUE

| | | |
|---|---|---|
| | | |
| | | |
| | | |
| | | |
| | | |
| | | |
| | | |
| | | |
| | | |
| | | |
| | | |
| | | |
| | | |
| | | |
| | | |
| | | |
| | | |
| | | |

DEMCO 38-296

# PLANET UNDER STRESS

# PLANET
# UNDER
# STRESS

## THE CHALLENGE OF GLOBAL CHANGE

Edited by

**Constance Mungall & Digby J. McLaren**

**For the Royal Society of Canada**

Toronto   Oxford   New York
OXFORD UNIVERSITY PRESS

ynford Drive, Don Mills, Ontario M3C 1J9

Oxford   New York

ta   Madras   Karachi   Petaling Jaya

Tokyo   Nairobi   Dar es Salaam

Cape Town   Melbourne   Auckland

and associated companies in
Berlin   Ibadan

GRAPHICS BY STEVE CARISSE,
CARISSE GRAPHIC DESIGN LTD, OTTAWA

REPRINTED WITH CORRECTIONS 1991

Canadian Cataloguing in Publication Data
Main entry under title:

Planet under stress: the challenge of global change

Includes bibliographical references.
ISBN 0-19-540731-8

1. Biosphere. 2. Man—Influence on nature.
3. Ecology. I. Mungall, Constance. II. McLaren,
D. J. (Digby J.). III. Royal Society of Canada.

QH541.P53 1990        33.95'16        C90-093400-X

OXFORD is a trademark of Oxford University Press

345–4321

Printed in Canada

# Contents

Acknowledgements

x

Preface

DIGBY J. McLAREN

xiii

PART I
*Our Planet Observed:
The Assault by* Homo Sapiens

WILLIAM S. FYFE

*Population* 3
*Food* 5
*The Planet Observed—1* 7
*The Planet Observed—2* 9
*Soil—1* 11
*Soil—2* 13
*Our Planet's Energy* 15
*Warming?* 17
*The Atmosphere* 19
*Waste* 21
*Pollution—Air* 23
*Water* 25
*To Manage a Planet* 27

PART II

*Setting the Stage*
29

CHAPTER 1

Dynamics of Planet Earth
WILLIAM S. FYFE
30

CHAPTER 2

The Changing Atmosphere
GORDON A. McKAY and HENRY HENGEVELD
46

*Policy/Action Recommendations*
68

*What Can I Do to Protect the Atmosphere?*
70

*El Niño*
BEV McBRIDE
71

*How Is Human Health Endangered?*
CLAIRE A. FRANKLIN
73

CHAPTER 3

Our Fragile Inheritance
W. RICHARD PELTIER
80

CHAPTER 4

Approaching Today
JOHN V. MATTHEWS JR
96

*The Last Glaciation: The Laurentide Ice Sheet*
ARTHUR S. DYKE
110

*Ice Cores*
R. KOERNER
112

PART III
*A Closer Look*
115

CHAPTER 5
The Polar Regions
FRED ROOTS
116

CHAPTER 6
The Tempering Seas
LAWRENCE A. MYSAK and CHARLES A. LIN
134

CHAPTER 7
Fresh Waters in Cycle
D.W. SCHINDLER and S.E. BAYLEY
149

CHAPTER 8
Forests:
Barometers of Environment and Economy
J.S. MAINI
168

*Brazil*
185

CHAPTER 9
Grasslands into Deserts?
JOHN STEWART and HOLM TIESSEN
188

*World Conservation Strategy*
SUSAN HOLTZ
207

PART IV

*Managing Ourselves as Part of One World*
209

CHAPTER 10

Grounds for Concern:
Environmental Ethics in the
Face of Global Change
PETER TIMMERMAN
211

*Changing the Globe, Changing our Minds:*
*Towards a Global Contract*
IAN BURTON and PETER TIMMERMAN
220

CHAPTER 11

People Pressure
SUSAN A. McDANIEL
225

*The Demography of Development*
NATHAN KEYFITZ
239

*Family Planning in Zimbabwe*
242

CHAPTER 12

From Technological Fix to
Appropriate Technology
DONNA SMYTH
244

*Satellite Remote Sensing*
JOSEF CIHLAR
263

*Reflections on Science and the Citizen*
URSULA M. FRANKLIN
267

CHAPTER 13
## The Missing Tools
ROBERT GOODLAND and HERMAN DALY
269

*The Hamburger Connection*
282

*Industry's Response*
ROY AITKEN
283

CHAPTER 14
## Surprise and Opportunity:
## in Evolution, in Ecosystems, in Society
C.S. HOLLING and STEPHEN BOCKING
285

CHAPTER 15
## Peace, Security, and New Forms of
## International Governance
FEN OSLER HAMPSON
301

*The Challenge of Environmental Peacemaking*
GERALDINE KENNEY-WALLACE
318

CHAPTER 16
## Summing It Up
J. STAN ROWE
322

## For Further Reading
333

## Contributors
338

## Index
341

# Acknowledgements

From conception to publication this book, which came to be known as 'the global change book', had a gestation period of over three years. Many people were involved in that time on many levels, and the Royal Society wishes to thank them all. The president, Digby J. McLaren, then president-elect, conceived the idea of bringing the concerns of the International Geosphere-Biosphere Program to the general public. He maintained direction throughout the project, which coincided with an expansion in the activities of the Society. He early enlisted an editor, Constance Mungall, who had long experience in making scientific research accessible to lay readers. Her assertive and tenacious influence shaped the final product. They were joined as an editorial committee by Michael Dence, the executive director of the Royal Society; Glenn C. Sutter, the co-ordinator of the Canadian Global Change Program; John Fyles, of the Geological Survey of Canada; and J. Stan Rowe, professor emeritus at the University of Saskatchewan. Of these, the last two contributed most abundantly and significantly. John Fyles, who had served in an advisory capacity to the Berger Commission and the two-volume report that followed it, lent his wisdom and scientific knowledge throughout the evolution of the book. And Stan Rowe, the author of the last chapter, 'Summing It Up', of necessity and generously read every preceding chapter, offering guidance, as the book progressed, from his knowledge of ecology and of life, and commenting helpfully in his overview.

Along with the editorial committee, the thirty-three authors of chapters and boxes participated in a process that itself was an element in the shift in thinking necessary to deal with global changes. Not only did they take part in authors' workshops to share ideas with other researchers in a range of disciplines, but they faced new ways of communicating and relating to a readership most of them, in their academic orientation, did not usually address. We wish to thank all the authors listed in the biographies at the end of the book, but one, William S. Fyfe, dean of science at the University of Western Ontario in London, deserves special mention. He wrote, often on aeroplane trips between international scientific meetings, not only Chapter 1, but the pictorial introduction to the whole book and the introductions to sections. His insights, his experience in speaking to lay audiences, and his humour were often available to the editorial committee.

Each chapter, in addition, was read by an expert in the field discussed. The reviewers included Darwin Anderson, University of Saskatchewan, Saskatoon; Marie Campbell, Ottawa; John Dauvergne, Industry, Science and Technology Canada, Ottawa; Lynda Dredge, Geological Survey of Canada, Ottawa; Sylvia Edlund, Geological Survey of Canada, Ottawa; Doug Heyland, Science Institute of the Northwest Territories, Yellowknife; Paul Keddy, University of Ottawa; Alan Longhurst, Bedford Institute of Oceanography, Dartmouth; Gray Merriam, Carleton University, Ottawa; Dorothy Smith, Ontario Institute for Studies in Education, Toronto; Robert Stewart, Agriculture Canada, Ottawa; Richard Stoddard, Fisheries and Oceans Canada, Ottawa; John Tener, Ottawa.

Graphics were executed by Carisse Graphic Design Ltd, with consultation by Jean Geuer.

Other people contributed in unique ways, entailing time, energy, and creativity. Three fellows of the Royal Society stand out in this category. The first is Hugh Wynne-Edwards—at that time vice-president research and development of Alcan Inc., Montreal, and now engaged in private industry and research in Vancouver—who advised about the project from the beginning and participated in authors' workshops. Another was Gilles Paquet, fellow in residence at the Institute for Research into Public Policy, Ottawa, who helped to restructure the book so as to integrate the chapters on the social sciences. Finally, John M. Robson, professor of English at Victoria College, University of Toronto, and honorary editor of the Royal Society, read all the chapters, making revealing and useful comments about logic and grammar.

Many others made valuable contributions along the way. They include Elise Boulding, Boulder, Colorado; Linda Demers, Canadian International Development Agency, Ottawa; Margaret Eichler, Ontario Institute for Studies in Education, Toronto; George Hobson, Polar Continental Shelf Project, Ottawa; Barclay McMillan, MetAbility, Ottawa; Alan Morgan, University of Waterloo; Alex Mungall, Ottawa; Ruben Nelson, Canmore, Alberta; David Peat, Ottawa; Gail Stewart, Ottawa; Jan Veizer, Universities of Ottawa and the Ruhr; Marjorie Wilson, Annandale, Va.; US National Aeronautics and Space Administration, Washington, D.C.

The Society is particularly grateful for financial contributions from:

Natural Sciences and Engineering Research Council of Canada
Social Sciences and Humanities Research Council of Canada
Secretary of State
Energy, Mines and Resources Canada: Surveys, Mapping and
    Remote Sensing Sector
Energy, Mines and Resources Canada: Geological Survey of Canada
Environment Canada
Forestry Canada
Canadian International Development Agency
Canadian Institute for International Peace and Security
Pan Canadian Petroleum Ltd.
Shell Canada Ltd.
Unocal Canada Ltd.

The book was groomed by William Toye, the Editorial Director of Oxford University Press Canada, who steeped himself in an unfamiliar field to give both textual and illustrative material his characteristic care.

THE ROYAL SOCIETY OF CANADA

# Preface

### DIGBY J. McLAREN

*President, The Royal Society of Canada*

We now see the Earth as a small planet in space that is inherently changeable. Life has played an important role in shaping the physical and chemical nature of the planetary surface. But life developed in balance with the changing environment as a result of an evolutionary process driven by those changes. In the very recent past, the emergence of the human race began to cause change in the environmental flux more rapidly than, and in a different manner from, the established system. With essentially free energy supplied by fossil fuels, our race has become, during the last two centuries, a dominant force for change on Earth.

This book is the result of bringing together experts from many fields and persuading them to tell us what they have learned about the current changes. This is not an official version, arrived at by consensus or coercion. Although it inevitably makes some generalizations, it does not attempt superficial conclusions. The Royal Society of Canada acted as catalyst and co-ordinator in this endeavour, and provided an editor to help the communication of scientist to lay person, but it sought no uniform view and imposed no opinion. We offer, therefore, a body of factual and informed description interpreted by a group of authors, mostly scientists, who give you the best explanations they can and suggest various future scenarios for the fate of the Earth and its inhabitants. Many disagree among themselves in small matters and even in some large matters, but you are witnessing a new process: scientists in many disciplines, sociologists, demographers, philosophers and moralists, interacting and attempting to speak a common language of concern for the health of the planet. This is an unusual phenomenon. We hope that these essential interactions have not been too long delayed for us to take measures globally to head off a developing disaster.

The Royal Society of Canada began to focus its interest in Global

Change in 1984. It was represented at Berne in September 1986, when the International Geosphere-Biosphere Program was initiated by the International Council of Scientific Unions. It soon became apparent that because of the urgency of the problems, we must seek understanding about the changing dynamics of the Earth from all disciplines of knowledge: the natural sciences, the social sciences, and the humanities. The Royal Society also realized that perhaps the most important task was to inform the public as fully as possible on current research into global change, on the significance of the results as they are currently being interpreted, and on the social, economic, legal, and moral aspects involved in coming to terms with the most serious threat ever to have arisen to the healthy functioning of the planet, the home and support of humankind.

The human being is an animal that has moved out of ecological balance with its environment. Humankind is a wasteful killer and a despoiler of other life on the planet. This normal and apparently acceptable behaviour has been licensed by a belief that our use of the Earth's resources is God-given, and encouraged by an economic system that emphasizes short-term profit as a benefit. We are only slowly learning to put a real cost on the resources we consume, and the wastes we produce. Humankind is now dominant in effecting perhaps irreversible change on the Earth's surface, and I suggest that we do not know enough to decide how to run this planet. We are forcing our will upon it, using ever-depleting resources and increasing waste discharge, while at the same time claiming that we must aim for 'sustainable development', which has been defined as development that meets the needs of the present without compromising the ability of future generations to meet their own needs. This forces us to consider if the present level of resource use and waste production can be sustained, to question the concept of growth economics, and to admit that it would be impossible to extend current Western standards of living to the world as a whole.

It seems to me that there are two major forces at work preventing us from attaining true global sustainability:

(1) Population growth seems to be the single most important factor in increasing environmental stress, including depletion of materials and energy resources and a runaway increase in solid, liquid, gaseous, and heat waste. We are currently adding one billion to the population of the world every eleven years. Most of the resources and most of the waste produced are due to the activities of a relatively small proportion of the total population. In certain areas of the world people are reproducing much more rapidly than in the West, but a Western baby will drive a car when it grows up, and will use resources and produce waste a hundred to five hundred times more than many

of the babies being born in the Third World. Recent studies suggest that with current dietary patterns we have reached or even exceeded the carrying capacity of the globe in population.

(2) Our use of fossil fuels continues to grow. If the relative decline in the use of petroleum products as fuel continues, it will be more than offset by an increase in coal-burning, which will further increase the greenhouse gases and particulates in the atmosphere. That in turn will increase the warming effect, which appears to have started. Bearing in mind the inevitable change in climate and rise in sea level that results from warming, how much can we accept?

Scientists can point out and measure many of the changes taking place in our immediate environment and our own lives. But it will take very much more than science to change our current system, and persuade us to learn to live in balance with Earth's ecosystem. All people on Earth are in this together, and so must find a joint solution. It is too late to build walls around, or put roofs over, regions of the world. The problems are exclusively global, and the solutions must be also. We are now faced with a task that is more difficult than anything we have ever contemplated: to decide how we may continue to live on this small planet. Any departure from ecological balance that destroys most of the remaining life on earth—and the big killing is under way—will mean that we are doomed to a similar fate. In other words, it is imperative that we learn to live in balance with all of Earth's complexities. Humankind has never before been faced so urgently with such a challenge—but face it we must, or life on this planet, for human beings, may become insupportable.

PART I

# Our Planet Observed:
# The Assault by Homo sapiens

WILLIAM S. FYFE

*Copacabana Beach, Rio de Janiero. (Courtesy William S. Fyfe.)*

# Population

*Homo sapiens* is one among over ten million species living on the planet Earth. When Christ was born there were 200 million of us. This number had doubled to 400 million 1,500 years later. The last doubling (1950-87) happened in thirty-seven years, bringing our population to its present number of over five billion.

Population growth in the 'rich' countries is almost level, or is declining. Some countries, like China and Indonesia, have declared that the current rate of growth cannot be sustained, and have instituted successful population-control measures. However, in many countries families are still large. For example, in Kenya 8 children is the average; in Egypt, 5; Nigeria, 6.6; India, 4.3. In Kenya the population grows at 3.9 per cent per year, in Nigeria 2.8 per cent per year, in India 2.3 per cent per year. Nations that have high birth-rates seem also to be the most poverty-stricken. But the picture is not quite so straightforward: Saudi Arabia, for instance, with an average of 6.9 children per family, is not poor.

Today we add 90 million people per year, the equivalent of the population of Canada each four months, of the USA each two-and-a-half years. At this rate the global population of *Homo sapiens* will reach ten billion before the year 2050. It is doubtful that the planet will support such growth without ever-increasing distress. Ultimately population is controlled by the basic resources of food and water. An indication of the growing strains in the system is the increasing numbers of environmental refugees—ten million people at the present estimate—who must move because their systems for support are collapsing.

Population growth is a very complex issue, influenced by social-religious-educational-political structures, and particularly by diverse views of the world, the place of all people in it—and perhaps by the status of women. There is little doubt that the present flood of environmental problems is related to population growth. We are recognizing the difference between growth and sustainable growth.

*Women and children drought victims in the Sudan wait in line for daily rations. (Courtesy Reuters/Bettmann Newsphotos.)*

# Food

Few of us in the developed (rich) countries know, or have felt, the meaning of the word hunger. Few of us worry about tomorrow's breakfast. We eat a large variety of food: salads, meat, fish or eggs, fruit, bread, milk, juices . . . . In a normal diet we may eat a dozen food products daily, many imported over large distances. Our longevity has been greater, our health never better.

But for something like sixty per cent of the world's population hunger is a constant companion, largely associated with rapid growth of population, environmental deterioration, and social disruption. In the search for food, people are forced to move, and often this move is to the cities.

At present over 40,000 children between the years of one and five die each day, usually from malnutrition or water-borne diseases. The situation is growing worse. People in Brazil, for example, have less food per person every year, a situation that has plagued Africa for decades.

Food, of course, does not automatically mean nutrition; these are two quite different things. Good nutrition normally means food diverse enough to supply the complex spectrum of the nutrients we need for good health and for a good immune system to fight off the bacteria and viruses and other organisms that may make us sick, or even kill us.

Recently the United Nations Population Fund published a report, 'The State of World Population 1989'. In the discussion of nutrition we read that undernourished women in Bangladesh lose forty-three per cent of their children by miscarriage or infant death. In India, where in general people get the necessary calories, sixty-eight per cent of the population is nevertheless anaemic, and in Egypt seventy-five per cent. Malnutrition leads to a host of problems: stunted growth, rickets, learning difficulties, risk of diarrhoea, bone deformity, goitre, and many infectious diseases. And malnourished people have very little energy.

For many of these people there is little or no variety, let alone a surplus. If the world population continues to grow at ninety million per year without some global food planning, it is certain that we will see catastrophic famines, and millions of children of malnourished mothers will be deformed physically and mentally. The laws of nature—the demands of our physical systems—apply to *Homo sapiens* as to all other forms of life.

# The Planet Observed—1

In the middle of the twentieth century we saw our planet from space for the first time. Today we can see all of our planet, and how it is changing, in quite remarkable detail. There is no doubt that this new ability to observe is changing our idea of our relationship to the planet.

The image opposite—a mosaic of photographs taken from US weather satellites between 1974 and 1984—represents a North-to-South satellite transect from Europe to Africa at night. It shows magnificently the concentration of energy in the North as well as the social and geographical differences on our planet. In contrast to daytime images, where only natural features are easily visible, this mosaic of night-time images makes it possible to trace human technology.

Much of Africa is dark. The major sources of light are from controlled fires, the result of grassland burning, slash-and-burn agriculture, and the clearing of forests. These fires are prominent throughout the highlands of the sub-Sahara savannas and East Africa. The giant bright spots in the Persian Gulf are also from burning, the flaring-off of natural gas from the great oil-gas fields—an image of waste. Ironically we see the same phenomenon in the Niger Delta region of energy-poor Nigeria.

The North is alight with the developed technology that provides its population with a high material standard of living. In the energy- and technology-poor South, poverty and hunger abound.

The populations of the two segments, north and south of the Mediterranean, are about equal. But the population of the North is not growing. In the South—where food per capita is declining—the population is growing about three per cent a year.

*A North-to-South satellite transect from Europe to Africa at night. (Courtesy US National Aeronautics and Space Administration.)*

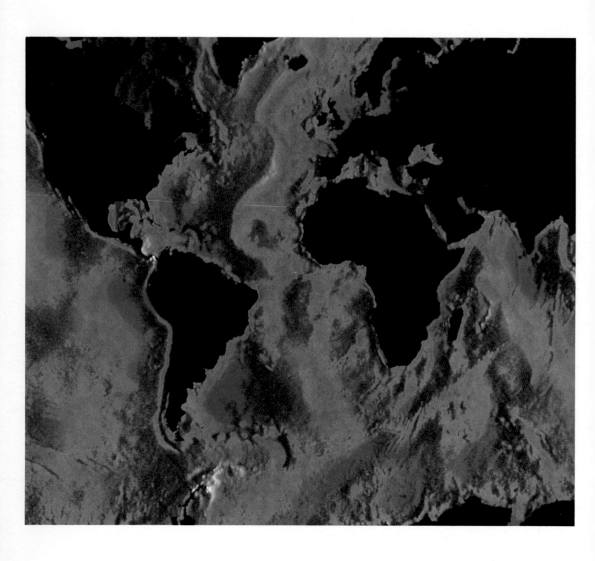

# The Planet Observed—2

This remarkable relief map of the topography of the ocean floors was produced from an analysis of the variations in the distance between an orbiting satellite, flown by NASA in 1978, and the ocean surface. It means that we can find all major features of the ocean floor from a satellite in only a few months. The analysis depends on Newton's classic laws of gravitational attraction: the sea surface reflects the distribution of mass in the water column and the solid earth beneath the water.

This picture reveals many of the features that prove our planet to be convecting—that is, hot, light material flows towards the surface, where it cools, loses gases, becomes heavier, and flows down again to deeper levels. This process gives us the volcanic hot spots that form island chains like those of the Hawaiian system; the long ocean ridges like those in the mid-Atlantic, where energy from the Earth's interior moves to the surface, and the great ocean trenches near the Pacific margins where the ocean floor flows back to the Earth's interior.

Such observations of Earth and other planets in the solar system will lead to better understanding of how planets form and change with time. They are essential in planning sustainable development for our future.

*Map of mean sea level recorded by spacecraft altimeter in 1978, indirectly revealing the topography of the ocean floor. (Courtesy US National Aeronautics and Space Administration.)*

*Desertification happens as a result of drought or overuse, and one exacerbates the other—as this scene in Cameroon, West Africa, shows. (Courtesy Roger Lemoyne, CIDA/ACDI.)*

# Soil—1

The nutrient supply for more than five billion human beings comes almost exclusively from plants and animals that live on the surface, on and in the stuff we call soil. (Globally, we depend very little on food from the sea.) Soil forms when organisms, water, air, and surface materials interact. It provides the physical support for the plant, and the nutrient and water bank from which the plant draws its sustenance. It carries a load of minute micro-organisms—bacteria, fungi, viruses, insects, worms—that play a vital role in supplying nutrients to plants—and without a certain soil thickness, growth of most of our food crops would be impossible. It takes a few thousand years to produce a useful amount from rock; but in dry regions that have little rainfall, like the Canadian prairie, inches of soil can be lost in a single rain or wind storm. Soil that is washed away collects in waterways and affects navigation, irrigation, and even the generation of electricity as the reservoirs behind dams fill with silt. Eventually most of the lost soil forms sediment in the oceans.

Across the world, as agriculture becomes more intensive, as more and more herbicides, pesticides, and fertilizers are used, and as the plants in a field become less diverse, land becomes more vulnerable to erosion. Desertification—the making of more desert—occurs when vegetation growth is reduced in semi-arid regions owing to droughts or less-than-usual rainfall; or as a result of overuse (too much cultivation and grazing). The deserts of North Africa have been spreading across the continent for decades. Drought has forced local people to cut all the trees to provide cooking fuel and to feed animals. The photograph opposite exemplifies the barren state such areas are reduced to. In another part of the world, sand from the great Gobi Desert is constantly blowing away, much of it into the city of Beijing 150 miles away. Areas threatened by desertification represent thirty-five per cent of the world's land area, affecting nineteen per cent of the world's population.

# Soil—2

Pressure on existing arable land, the need to develop more to produce food and cash crops, and the constant and overwhelming need for paper (for each issue of the Sunday *New York Times* an estimated 77,000 trees are felled) lead to the destruction of whole forests; and the invention of the chain-saw allows us to pursue this destruction with ruthless efficiency. As population continues to expand, all the world's forests are threatened.

One result of deforestation is vividly demonstrated in this satellite picture of the mouth of the Betsiboka River in Madagascar. A forest has been demolished. It is no longer absorbing water, or holding the soil in place on the hillsides, so the rivers flood and carry away the topsoil to the sea. In the tropics particularly—where most of the world's deforestation is taking place—the soils are quickly lost after the clearance of a forest, which is unlikely ever to be restored.

Forest terrains, both in tropical and cooler climates, play a vital role in the global system. When rain falls on a forest, as much as eighty per cent may be returned to the atmosphere by evaporation through the leaf systems. This return of moisture keeps the atmosphere from drying out and reduces stream erosion. Photosynthesis in the leaf systems uses carbon dioxide from the atmosphere to produce carbon compounds to build plant tissue. This process has for eons played a major role in maintaining the balance between carbon in living systems and carbon in the atmosphere. In other words, before humans entered the picture it controlled our climatic greenhouse. The greenhouse is now warming because of an increase of carbon dioxide in the atmosphere, resulting partly from deforestation. Forests store more carbon than any other vegetation on land. In 1987, 8 million hectares of Amazon rainforest were cleared by burning. In 1989, 6.4 million hectares of Canadian forest burnt, partly as a result of dry seasons.

*Streams of mud entering the Indian Ocean from the mouth of the Betsiboka River, Madagascar—the result of deforestation. (Courtesy US National Aeronautics and Space Administration.)*

*Deforestation in St Lucia, British West Indies. (Courtesy Hélène Tremblay, ACDI/CIDA.)*

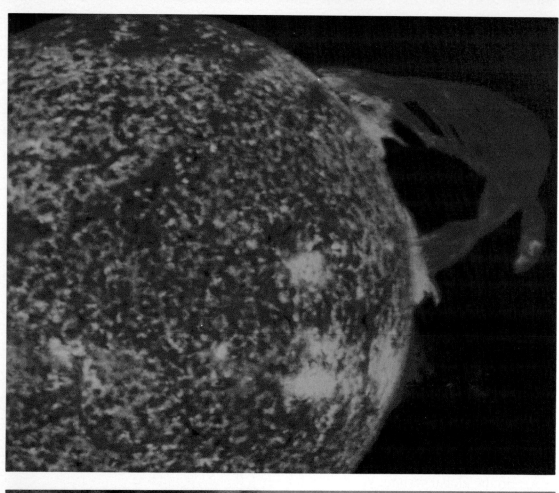

# Our Planet's Energy

Our unique planet, and our species, exist because of two great energy sources: the Sun and the Earth's hot interior.

During the day our Sun, a nuclear-fusion reactor, pours energy onto the surface of the Earth. Here plants and animals use this energy through the process of photosynthesis, which produces all our food, our forests, our life-support system—and perhaps our sanity-support system.

Energy from the Sun is not constant. The Sun is dynamic, and during solar flares or sunspot cycles the energy fluctuates. There are also major fluctuations due to the changes in Sun-Earth orbital motions. It is these changes that lead to phenomena such as the great Ice Ages.

The energy fluxes between the Sun and the outer layers of the Earth are critically dependent on the composition of our atmosphere. For four billion years at least, the waters on Earth have not boiled (like those on Venus) nor totally frozen (like those on Mars) because of our remarkable global thermostat: the balance of all the planetary processes, including the biological, that influence atmospheric chemistry.

In the deep Earth there is another reactor, this one using nuclear fission. The radioactive fuels—natural elements like uranium and thorium—produce the energy that erupts in volcanoes. The rise of hot materials from the interior builds the mountains that then modify wind and wave patterns and shape and move the continents. These deep thermal processes control the entire flow of nutrients into the living system because they form our surface topography, which in turn controls the flow of surface waters. Without such processes, the Earth would soon become a great dismal swamp.

So all are linked: Sun, Earth, Life.

*A photograph of the sun from NASA's Skylab 4, showing a spectacular solar flare. (Courtesy US National Aeronautics and Space Administration.)*

*Hawaiian volcano erupting. (Courtesy Lyn Topinka, US Geological Survey.)*

# Warming?

Global temperatures are rising—observations made over the past two to three hundred years shows us that. And yet the energy from the Sun has changed little, and has even decreased from 1980 to 1986. Are we seeing the influence of the changing atmosphere, the greenhouse or warming effect caused by an increase in gases like carbon dioxide and methane in the atmosphere?

The sea level is rising as the ice stored in massive glaciers diminishes—another result of the greenhouse effect. The satellite picture opposite shows the great Byrd glacier of the Antarctic continent flowing to the sea as a viscous river of ice. This flow feeds the huge ice sheets of the continental edge, which in turn may fragment into icebergs and then melt.

If these rivers of ice were to flow a little faster, and if all this ice melted, the sea level would rise about 60 metres, flooding more than half the great cities of the world.

Ice at and near the polar regions plays a vital role in the global climate. When the polar oceans freeze, the water becomes more salty and heavy. It sinks, driving deep, cold ocean currents away from the poles, and sweeping away dissolved carbon dioxide to be used by marine organisms further south. Ice sheets also reflect light from the Sun. Without this ice cover, the polar oceans would rapidly warm and the major ocean-current patterns would change, along with global climate. New data from the Arctic shows that the sea ice near the pole is thinning rapidly, as much as forty per cent since 1976. If this continues there could be a very rapid change in northern climate. Remember what happens to the temperature of your drink when the ice cubes melt!

With satellites we can now monitor such changes on a global scale.

*Byrd glacier, Antarctica, flowing through the Transantarctic Mountains into the Ross Ice Shelf, as seen from space. (Courtesy US National Aeronautics and Space Administration.)*

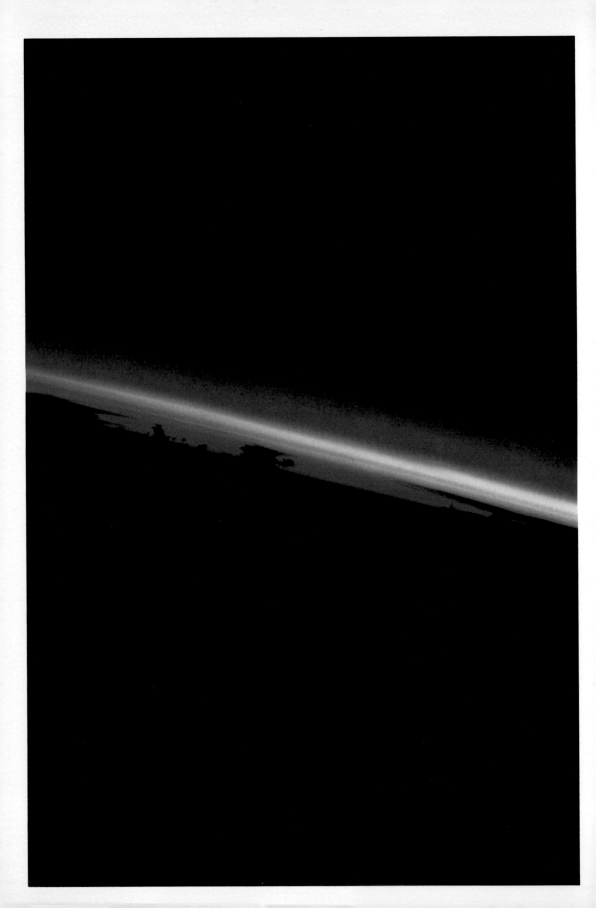

# The Atmosphere

This remarkable photograph from a space shuttle shows the interface between the solid Earth and the dark void of space that contains little matter to scatter light. The dark jagged clouds mark the boundary of the troposphere in which we live. The pinkish-white layer above—the stratosphere—is 15 to 30 kilometres high and shows sunlight scattered by fine particles. This stratospheric layer contains ozone, a molecule made up of three oxygen atoms ($O_3$), which absorbs most of the lethal ultraviolet radiation coming from the Sun. Without this shield, life would not exist on the surface of the Earth: ultraviolet radiation sterilizes all forms of life familiar to us.

In the last two decades scientists have noticed that the concentrations of ozone have been declining. They think this is due to the use of chlorofluorocarbons (CFCs), chemicals used in aerosol sprays, refrigeration, dry-cleaning, and plastic foam as in hamburger cartons. These CFCs break down ozone.

The ozone layer over the Antarctic is at its lowest concentration in September and October. Satellite pictures show us that the 'ozone hole' over Antarctica is spreading. There is growing evidence that the ozone layer over the Arctic is shrinking. Again, CFCs seem to be responsible.

Our atmosphere is one of our most vital resources and must be protected.

*The stratospheric layer, containing ozone, which absorbs and shields us from much of the ultraviolet radiation coming from the sun. (Courtesy US National Aeronautics and Space Administration.)*

*Living with garbage in Cairo. (Courtesy of Chris Baker, London, England.)*

# Waste

Cairo's garbage and sewage systems were designed for a city of one million. The population has reached twelve million and the city can treat only half its sewage. Polluted water is the major cause of death of 131 Egyptian children in every thousand who die before the age of five. Cairo is not unique: this is a typical scene in the new Third World megacities.

Until recently we have assumed that our environment is so vast that it can easily accommodate the dispersion and dilution of all noxious properties. But we are now so numerous, and our material needs are so complex, that this concept is no longer valid.

Every Canadian produces three kilograms of waste per day—Canada leads the world! Even in less-developed countries, people produce one kilogram per day. Our garbage contains not only newspapers, which can be recycled, but many non-biodegradable things like plastic, old film, cadmium batteries, poisonous drugs—much of it deadly to life. When we dump untreated sewage into our waters we create floods of toxic organisms that can toxify and destroy fish.

We have only temporary solutions for dealing with waste—what to do with the mountains of garbage we produce except bury it? The problems grow larger as more and more people move into the great megacities that blight the world landscapes, cities of twenty million and more. Almost half the people on Earth draw their water from wells in the subsurface. When we bury wastes—unseen, the effluents may pollute this vital resource.

It becomes very clear that one of the best ways to deal with garbage is to produce much less of it. Stop wrapping everything in multilayers of plastic. Stop, in this age of electronic communications, using so much paper. Reduce the harmful influences of our disposable products by separating waste into its major components. Recycle metal, glass, plastics, paper. Return animal-vegetable waste, properly treated, to farmland as organic fertilizers.

# Pollution—Air

Human activities, in combination with modern technologies, are producing a vast number of new and highly specialized chemicals and chemical species that change our soils, atmosphere, and water systems. Fossil-fuel combustion pours sulphur and nitrogen acids into the atmosphere. Humankind disperses into the environment more than 50,000 compounds that do not occur naturally on the planet. We could call them zenobiotic—foreign to life.

Acid rain is one of the most destructive pollutants in Europe and North America—indeed, wherever there are large cities that have high concentrations of industry and automobiles. The burning of fossil fuels produces waste gases such as sulphur dioxide and oxides of nitrogen, which are converted in the atmosphere to produce sulphuric acid and nitric acid and have a damaging effect on lakes, streams, groundwater, forests, agriculture, buildings, and—when we inhale sulphur dioxide—on our human health. The Red Spruce in Quebec (top) is dying as a result of acid fog, which has an acid concentration ten times that of acid rain. The lower photograph shows a gas cloud, with acid rain, pouring from the great metal smelters of Sudbury, Ontario. This industry has now promised to virtually eliminate their pollution by the end of the century.

In 1948 nineteen countries (including Canada) signed an agreement to reduce their emissions of sulphur by 1993. The United States, Poland, the United Kingdom, and Spain—four of the largest emitters of sulphur—did not sign. Much of the UK's emissions are wafted across the North Sea to Scandinavia.

But we must note that not all news is bad news. Today we live longer than our ancestors. For almost all people, better food, medical services, education, social structures contribute to our longevity. But in the 1970s the Club of Rome predicted that pollution would lead to a reduction of longevity by fifty per cent in the next century. Its prediction may be correct, unless we head it off by cleaning up the air we breathe and the water we drink.

*Dying trees on Mount Sutton, Quebec. (Courtesy Environment Canada.)*

*High-level release dilutes pollutants locally but creates regional problems far away. (Courtesy Atmospheric Environment Service of Canada.)*

# Water

For many, water is a precious commodity—it is not spread evenly over the globe, and without water there would be no life. We draw upon three great sources of fresh water: lakes, rivers, and the much larger quantity stored in the pore spaces of rocks near the Earth's surface. In regions of low rainfall or remote from rivers and lakes, if subsurface water extraction exceeds the rate at which the water is replaced from the surface, people must go deeper to obtain it and it often becomes salty. Again—there are limits to growth.

It becomes increasingly important to protect all our sources of water from pollution. Near this pulp-and-paper factory in Canada—as in most other regions of massive industry—the frothy waters of the St Francis River in Quebec are loaded with chemicals and algae. Flowing to the tide and estuarine zones of the oceans, they render shell fish toxic: Danger—Do Not Eat the Fish! Raw sewage in Venice and Hong Kong produces floods of toxic algae. In California the phosphate-loaded waste-waters from irrigation systems present problems for disposal. Fluids seeping from landfills, mining dumps, and coal-strip mines, acid and loaded with toxic metals, breed their own special forms of bacteria and algae. But in almost all these cases we have the knowledge to reduce such careless behaviour. We can vastly improve our waste technologies—and we know that it must be done now.

*The St Francis River, Quebec, near the Domtar Pulp and Paper Mill. (Courtesy SSC/ Photo Centre/ASC.)*

*Children defecating on a polluted beach in Bangladesh. (Courtesy Claude Dupuis/ IDRC.)*

# To Manage a Planet

As population increases, and perhaps doubles again by the middle of the twenty-first century, it is inevitable that humankind will invade and modify almost all parts of the planet once called 'natural'. For example, we debate today about attacking the resources of Antarctica, about saving whales and elephants!

We are reducing genetic diversity on our planet at an ever-increasing rate. Some estimate that we are destroying about one per cent of species annually. The tropical rain-forest opposite, with its vast diversity of species living from the forest canopy to the ground surface, is a centre of species evolution. Even the banana trees, probably introduced by the natives, flourish. Contrast this with a monoculture controlled by herbicides and pesticides, such as a massive field of a single species of wheat, which can be wiped out by climate change—in a dry season its soil can blow away—or disease, depriving people of food and farmers of their livelihood. Conditions on our planet fluctuate on many time-scales, and in the diverse rain-forest something will survive and take advantage of the fluctuation, like the weeds in our lawn that replace the grass when the season is dry. Our planet has witnessed massive extinctions from natural causes, but life has rebounded in both new and old forms. Diversity may very well be the key to its stability. Indeed, biological diversity may enable us to respond to climatic catastrophe on Earth in a way that promises our survival. These considerations are basic to planning our future strategies of sustainable development.

We have at last come to recognize that there are limits to humankind's many disturbances and manipulations of the ecosystem. We must face the threat to genetic diversity by forest clearance. We must have clean water, clean air, clean soil, and cleaner technologies. Though our planet still contains many mysteries, its workings are no longer unfathomable. We have received messages about our future, many of which we understand. We ignore them at our peril.

*Vegetation in a tropical rainforest on the slopes of the Sierra Nevadas of Colombia. (Courtesy Alvaro Soto/ IFIAS/Human Dimensions of Global Change.)*

*Planet Earth. Photographs like this, taken from space since 1968, have made us aware of the vulnerability of our planet home. (Courtesy US National Aeronautics and Space Administration.)*

# PART II

# *Setting the Stage*

We live on a planet in the solar system, a very unique planet in that it is almost covered by liquid water that, to our knowledge, has never boiled or totally frozen. This requires a delicate balance between energy, which comes from the Sun's and Earth's interiors, and the dynamics of energy-storage mediated by our unique atmosphere.

In this part of our story we will examine the geologic record and the evidence for the way our planet has changed since the birth of the solar system some 4.5 billion years ago. We will see that change on our planet is coupled with change in the biosphere, the beautifully complex array of species that inhabit our surface. We will see that change can be gradual, or at times catastrophic.

Surely the knowledge of our past history must be considered when we plan the sustainable development for our future. This past record is preserved—in rocks, ice, tree-rings—and provides us with the 'archives' of our planet's history.

WILLIAM S. FYFE

# CHAPTER 1

# Dynamics of Planet Earth

## WILLIAM S. FYFE

The first pictures of the planet Earth that came back to us from space gave us a new view of our remarkable global home. Beneath wisps of cloud, very different from the harsh reds and oranges of the other planets, Earth shines with a distinct blue tinge. About seventy per cent of the surface is covered by water—more, if you include the frozen ice caps at both poles. The land surface is partly rich green, partly beige to brown, and some of the large pieces seem to match each other like a jigsaw puzzle. The surface is highly irregular, with deep depressions and mountain ranges as well as broad, relatively smooth regions both above and below sea level.

The new view that space pictures have given us since 1968 is reinforced by our sophisticated sensing instruments. They tell us that this planet has a most unusual atmosphere, with over twenty per cent of highly reactive oxygen, which combines readily with many other substances. There is a strong magnetic field. The distribution of mass—as, for example, in the great mountain ranges—is uneven, because some areas have moved very recently, within thousands of years. The balance of forces, the dynamics of this planet's operation, are unique.

Our new view of ourselves as part of a great and intricate system, with an orderly if complex pattern, comes none too soon. The balances of energy-air-water-rocks, and the living organisms, are fragile, and they are shifting. One element, living organisms, and one species of that element, humanity, are having trouble fitting their behaviour into the pattern. Within the last 300 years human activity has become the dominant change factor on the planet. This is changing the interactions, accelerating long-term changes, and introducing life-threatening hazards. Slowly awakening to our danger, and using the remarkable abilities to observe our own natural systems that we have developed in the twentieth century, we are beginning to search for more understanding. Before we can assess the global changes and learn how better to manage ourselves on our planet home, we need to learn more about the dynamics of the planet as they operate now. Let us then look at how our beautiful Earth maintains the balance we are threatening.

## THE ENERGY BALANCE

One of the fundamental factors in the description of any natural system is the source of power that drives it. Clearly for the planet Earth, energy from the Sun is a primary source, one that changes with the orbital motions of the Sun-Earth system and with alterations in the distance of one from the other—and with changes in the Sun itself. It is possible that these variations are what trigger many aspects of climatic change, and events like the ice ages. How little we know about our Sun and its changes! To measure its energy output and fluctuations precisely, we must get above our atmosphere and use satellites. Only since the late 1970s have we been able to gather accurate data. We have learned that in the period 1980 to 1986 the total solar energy coming to our planet diminished, but that today the energy input is slowly increasing.

The planet has a second source of power. Deep within the Earth, certain elements—in particular isotopes of uranium, thorium, and potassium—decay, liberating heat, which flows to the surface. The molten outer core of the Earth, in cooling, acts like a thermostat. Thus the energy from atomic fusion in the Sun, and from radioactive decay in the Earth's interior, together stoke our power systems. The nature of our planet, and its continuous changes, must reflect changes in these dominant power systems.

## THE COOLING PROCESS

The Sun and planets formed about four-and-a-half billion years ago from dispersed cosmic matter dominated by hydrogen. Because of gravity, this gas and dust rapidly compressed, generating a huge amount of energy, melting or vaporizing most of the material in the planets. Early Earth, many believe, was largely molten, and this explains why our geologic record started only after about 500 million years, when the surface was cool enough to allow preservation of a few fragments of solid crust. During this early hot phase of planetary evolution, gases like nitrogen, and carbon dioxide, and water, were expelled to form a heavy atmosphere. Rain—probably hot rain, well above 100 degrees centigrade—followed. With the beginning of this precipitation from the water-loaded early atmosphere, a solid crust many kilometres thick formed over the entire Earth. Oceans formed at least 3.9 billion years ago. Reactions with rocks cleansed the atmosphere of acid gases and carbon dioxide and formed the salts of the oceans.

Heavy materials, like liquid iron, settled rapidly to form the molten, electrically conductive core, in which moving masses of matter acted as a dynamo producing a magnetic field. It is this magnetic field that screens the Earth from particles in the upper atmosphere's solar wind, which otherwise would destroy most forms of surface life.

Any hot object can cool in two ways: passively by conduction of heat, and by convection, where hot, light material moves towards the surface. The size of the body, and the rate of temperature change, determine the relative importance of the two processes. If the body is large and hot, convection is likely; and just as with the bubbling soup pot on the stove, the temperature distribution across the surface varies depending on the places where hot materials rise and cooler ones sink. On Earth we can recognize such sites by, for example, volcanoes, where hot molten rock flows to the surface. From thousands of heat-flow measurements we have discovered the fundamental patterns of convective motion, and know that about half the Earth's internal energy today is carried to the surface by convection, and about half by conduction. For the early planet, convection must have been the more important cooling process, heat being dissipated as the surface was folded in (Figure 1). It is these convection patterns that modify the surface and dominate the dynamics of our environment.

The exploration of the ocean floors has given us a large part of this story, a part missing before this century. Today, with satellites, we can measure the topography of the ocean floor for the entire planet; we can see it all.

## OCEAN RIDGES

The greatest and longest mountain ranges on Earth lie under the oceans. They include among others the Atlantic Ridge, the East Pacific Rise, and the Indian Ocean Ridge. Some of these volcanic ranges extend unbroken for 20,000 kilometres, half the circumference of the Earth. About ninety per cent of the Earth's volcanism occurs in submarine environments, almost all of it forming along the axes of these great undersea mountain ridges. Almost half of the energy produced in the Earth's interior is swept up by convective flow in the deep mantle about 100 kilometres below the surface, and discharged at the ridge volcanoes. The signs of this great underground cycle are the outpouring of lavas, and the creation of the high mountains deep below the surface of the sea.

## Evolution of the Continents

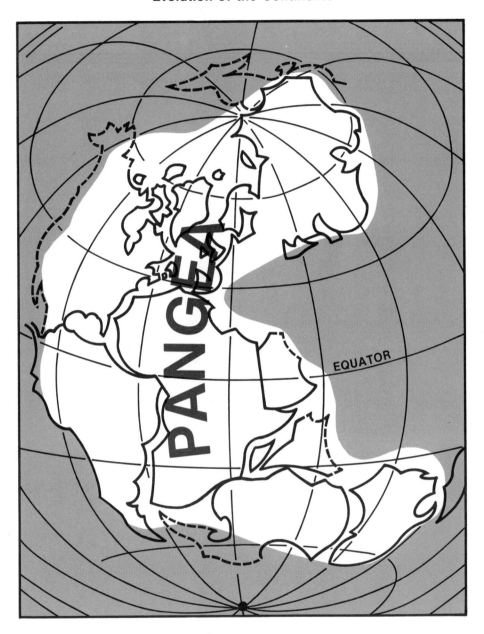

FIGURE 1. *Three hundred million years ago the continents we know today were fused together. Convective forces in the interior have moved the pieces into the present configuration. This former 'super' continent, Pangea, would have had very different climate patterns and been very harsh in the interior.*

*A vent in the East Pacific Ocean, about 2500 metres under the sea, belches hot mineral-laden water. (Courtesy Le Centre National pour l' Exploitation des Océans (CNEXO).)*

*Pillow lavas pour out near the ocean ridges and cover the sea floor, for example in the Sultanate of Oman on the Persian Gulf.*

The heat flows in erratic patterns, with some high temperatures, which we would expect where hot lavas pour out on the surface and are injected beneath the thin crust; but some temperaturs are low. Why the erratic behaviour? The most common lava is basalt, and when it crystallizes from the liquid, it contracts and produces a highly cracked and porous mixture. Water enters this cracked system: where it flows in, the surface is cold; where it is discharged, it is hot. About half the energy is carried away by such water flows.

The entire ocean mass is processed through the ridges every few million years—the ridges form one of the great chemical processing sites of the Earth. In the cooling process the original ocean water loses some elements, which are deposited in the rocks, producing mineral-rich basalts. When the water is discharged, they gain others. This exchange significantly influences the chemistry of the oceans and the nutrient supply to marine organisms. Like all volcanic processes, it is not steady but fluctuates, and we are slowly beginning to appreciate just how huge these fluctuations are.

The injection of new hot rock into the ocean ridges pushes the ocean floor aside, in a process we call ocean-floor spreading. It is this

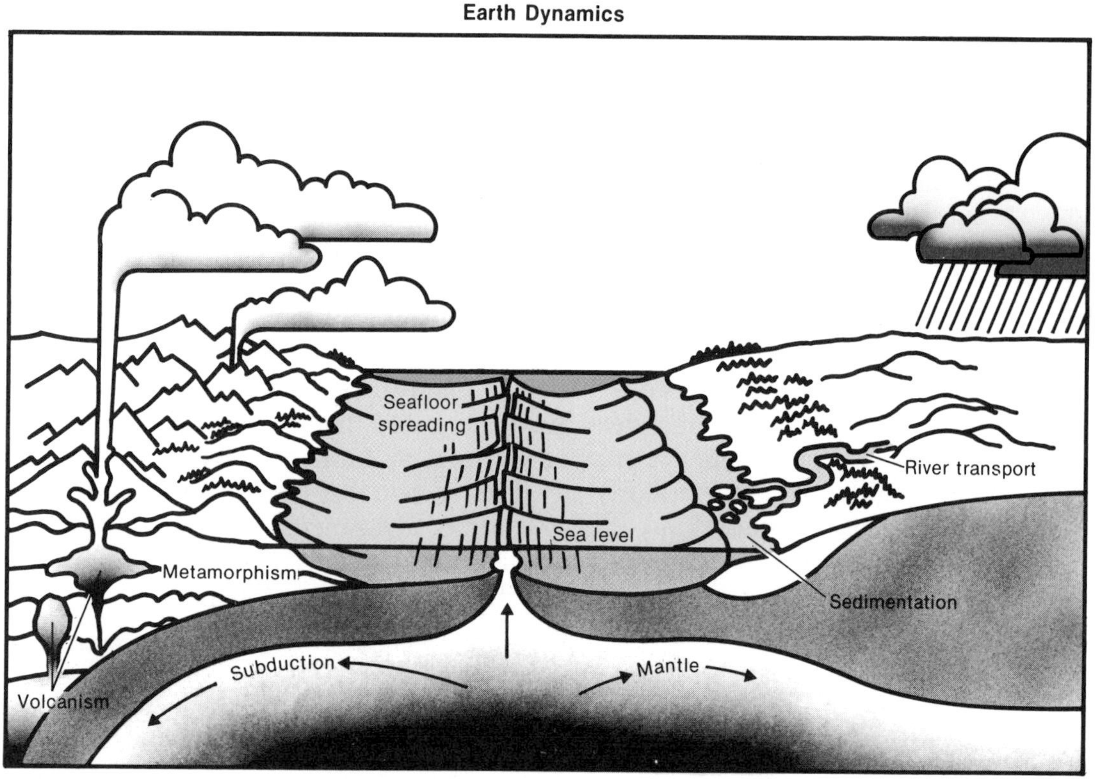

**Earth Dynamics**

FIGURE 2. *The Earth's convection today. Rising hot material builds the great submarine mountain ranges or ridges, pushing and spreading the continental masses. The cold ocean crust sinks back into the deep Earth, at least 700 km down. Explosive continental volcanoes and mountain ranges like the Andes rise up above these zones, as in the Pacific Ring of Fire.*

that powers continental drift—the ponderous journey of the continental plates across the Earth's surface, at about two to ten centimetres a year (Figure 2).

When basalt magma crystallizes, a minor product is the mineral magnetite, which becomes magnetized when it is grown in a magnetic field. When each new hot molten batch crystallizes, it takes up a magnetism related to the Earth's magnetic field, both in direction and intensity. Studies of the sea floor and of some terrestrial volcanoes show that periodically the magnetic directions reverse, implying therefore that the magnetic field of the Earth must reverse. The spreading ocean floor acts as a recorder of the oscillations in the magnetic field, which are in turn correlated with a definite age across the world. A reversal can happen quite fast, within thousands of years, and during the reversal the field is reduced in intensity. We are still debating whether these periods of a weak magnetic field, when the solar wind influences the surface with charged particles, correspond to periods of increased mutation, when new genetic characteristics are introduced. Certainly many species of fish and microorganisms seem to have a magnetic sense, and would have been quite confused in these transition times. We are still not sure if human beings have a magnetic sense.

## HOT SPOTS

The laborious movement of masses of matter and transference of heat continues in a systematic, regular sequence in the ocean ridges. Nevertheless some turbulence does occur, and produces volcanoes at apparently random sites on the Earth. The Hawaiian Islands of the Pacific are typical, but there are hundreds of such hot spots across the globe. The sources of their thermal anomalies seem to lie even deeper in the mantle than the ocean ridge processes; some think they may even be rooted near the core-mantle boundary, 2900 kilometres deep. In all convecting systems, very hot systems are the most turbulent. If oceans or atmosphere are further warmed by human intervention—particularly by contributions of gases produced by industry and automobiles—convection patterns could become increasingly chaotic. The same would be true for volcanism on a hotter planet. Perhaps significantly, the height of waves in the Atlantic Ocean is increasing, and hurricane Gilbert in 1988 had the highest wind velocities ever recorded.

## DESCENDING CONVECTION CELLS

If hot material flows upwards at the ridges, there must be a return flow of colder materials. The major flows back into the deep earth

TOP: *Mt St Helens, in Washington State, was venting heat from deep in the Earth's mantle when it erupted in 1980. (USGS photo.)*

BOTTOM: *A volcanic mountain on the North Island of New Zealand. The lake sometimes boils.*

occur at the sites of the great ocean trenches, or when mountain ranges of Himalayan type are pushed up. In places like the Pacific trenches, the cold upper layers of the ocean floor, about 100 kilometres thick, descend into the mantle at a velocity similar to the ridge spreading rates, two to five centimetres a year. There is little old ocean floor on our planet. The descending layer we call the lithosphere creaks and groans as it slides under the continental plates, and generates many of the great earthquakes. Most of the ocean floor is swept along, including some deep-sea sediments. This process, called subduction, recharges the deep earth with volatile components like water, carbon dioxide, and sulphur compounds, which are pushed into the mantle by the inexorable recycling. As the old, heavy ocean floor descends into the much hotter mantle, and is heated, it degasses, and fluids stream back to the surface; if they did not, the oceans would migrate to the interior of the Earth.

Near the trenches, where the lithosphere descends, the heat flow is very low. The cold material absorbs much of the heat flowing from the interior. Why is it, then, that normally a few hundred kilometres from the trenches we find volcanoes like the famous Pacific Ring of Fire? This narrow belt encircling the Pacific Ocean includes seventy per cent of the world's active and dormant continental volcanoes. They are formed because during deep subduction water streams off the descending lithosphere into the warmer overlying mantle, 50 to 100 kilometres below the solid rock crust. Water injected into the thick, solid mantle weakens it, so that it begins to move. This movement eventually leads to volcanism, pushing heat out of the interior of the planet. About a tenth of the energy transfer on Earth occurs by this process.

Most of the subduction-related volcanoes occur near the edge of continents, for example down the west coast of North and South America. The rising warm mantle under the continental crust at times melts the base of the continents and pushes up great masses of igneous rock—giant bubbles of granite, each often as much as 500 cubic kilometres in volume, which we call plutons. As the hot rocks move near the surface and encounter the ground water there, they are water-cooled. Many of the world's famous hot springs, like Yellowstone in the US and Wairakei in New Zealand, are created in this way. The same processes transport metals, and form most of our copper, molybdenum, tungsten, tin, and silver deposits.

All volcanic phenomena fluctuate in intensity. It takes time to collect the energy to melt rocks, time for the light magmas to accumulate the volumes compelling them to flow upwards. When Mt St Helens erupted in Washington State in 1980, the volcanic ash was a little over a cubic kilometre in volume. When the Long Valley volcano in Eastern

California erupted 730,000 years ago, the volume of magma blown out was 600 cubic kilometres. The hot ash thickly covered a disaster area of 100,000 square kilometres! Volcanologists think that such single events can reach ten times the size of Long Valley, and inject huge dust plumes into the atmosphere to modify the world's climate for years.

Not just volcanoes, but all the processes I've described so far—energy balance, cooling, ocean-ridge spreading, convection currents—are the dynamics of our water-cooled planet. While we see these systems operating in a spectacular way near volcanoes, we must also remember that they are happening everywhere on our planet: in the

*As hot rocks move near the surface and encounter ground water, they are water-cooled, producing hot springs. These geysers, about 75 miles north of San Francisco, are one of the world's largest sources of geothermal energy, but they tap only about four per cent of the identified geothermal energy in the US. (USGS photo.)*

quiet ocean floors, on the continents, the deep flow of waters transfers heat and material in solution to the surface. These materials form part of the nutrient cycle that supports life. We must remember too that many of these phenomena depend on Earth's outer shell of liquid water, the hydrosphere, which would not exist without solar energy.

## BUILDING THE EARTH'S TOPOGRAPHY

But Earth is alive—building mountains by the processes associated with convection. The hot rock rising from inside the Earth builds the high topography of the ocean ridges and mountain ranges. The ocean crust slides under the continents, and if this moving ocean floor itself carries with it a continent, a collision may result. For instance, some 100 million years ago the continent of India lay deep in the Southern Hemisphere. An oceanic trench lay near the present Himalayas. India drifted north and eventually collided with Asia. Most of this continental crust was too light to be subducted or dragged into the mantle. Instead, India and Asia pushed into each other and built a huge high region of thick crust, about 3,000 by 1,500 kilometres in area, the Himalayas. The process is still going on, as residual India pushes north at about five centimetres a year. More than half of the original Indian continental land-mass has been modified in this process.

Such collisions have occurred many times during the great recycling that molds continents and leads to global changes. Continents break apart or fuse together, oceans open and close. The consequences are myriad. For instance, some scientists now think that the formation of the Himalayas had a part in setting the great monsoon weather patterns of Earth today. Thus a mountain-building event in one place can change the total world climatic regime and, as with the Himalayas, totally change the supply of nutrients that support life in the oceans. Convection in the deep Earth modifies the life-support system of the surface.

## LIFE

Micro-organisms existed from the beginning, and they were probably a complex array. We know this from the highly variable sedimentary environments in many different places, including the floors of the Black Sea, the Pacific Ocean, and of both oxygen-rich and anoxic lakes. Our observations tell us that there are bacteria that can modify their biosynthetic pathways to cope with drastic differences in the local environment. We suspect that all through the planet's evolution the biomass had an enormous influence on our atmosphere and aquatic environments, as it does at present.

Given the primordial Earth's more turbulent circumstances—with more erratic volcanism, more meteorite impacts, more hot springs, and more rapid fluctuations and possibly dramatic differences in atmospheric circulation patterns—we would expect great local variability. That early high degree of instability may account for the dominance of the most robust simple systems of life: micro-organisms.

There is no question that all life is influenced by the environment. But is the reverse true? Does life influence the environment? Recently the British scientist James Lovelock has proposed the Gaia hypothesis,

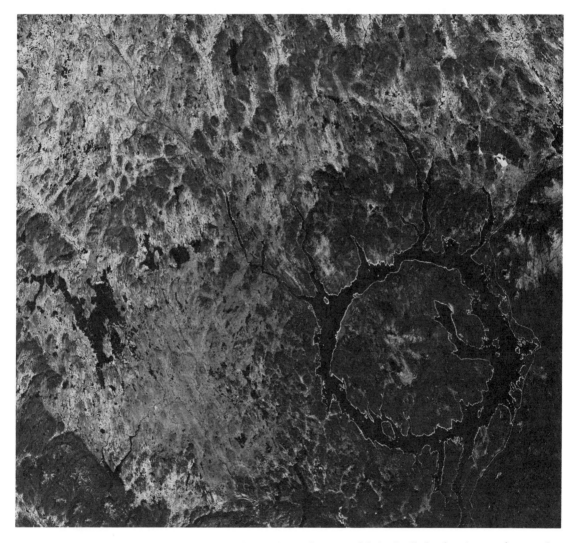

*During the evolution of the planet, bombardment from planetary debris rivalled volcanism as the most important geologic process shaping the Earth's surface. Manicouagan crater, Quebec, was scooped out 210 million years ago, and was originally 75 to 100 kilometres in diameter. (LANDSAT satellite image courtesy Canada Centre for Remote Sensing; Surveys, Mapping and Remote Sensing Sector, EMR Canada.)*

reinforcing the idea that life and the environment must be considered as a great unity.

We now have the tools to study this riddle. Modern molecular biology is developing ways to unravel life at the atomic-molecular level. And on the large scale, with satellites, we can watch the dynamics of life on our planet. We are beginning to understand and describe the complex interactions between geosphere and biosphere. Some features are clear. For example, our atmosphere, with about twenty per cent oxygen, could not exist without life and photosynthesis. The Earth contains a large mass of metallic iron, and when iron rusts it removes oxygen from the air. If this oxygen were not renewed by photosynthesis, we would have almost no oxygen left in the atmosphere. Again, the gases flowing to the surface of the Earth are simplified, carrying hydrogen, methane, and hydrogen sulphide; when carbon dioxide moves to the surface, living organisms reduce it to carbon and oxygen. Otherwise we would have a much higher carbon dioxide content in the atmosphere, and probably a warmer global temperature. Ice ages would never occur and the Earth would become like Venus, hot and barren. It is life on land and in the oceans that creates the balance, a balance that is critical to Earth, for carbon dioxide in the atmosphere forms part of our atmospheric thermal blanket—it largely controls the 'greenhouse' in which we live.


## PEOPLE

Once intelligent life evolved on this planet, a new force was introduced, eventually skewing the balance in a different way. Humankind has now become the dominant force in changing the surface of the Earth. In responding to population and industrial demands, we move rocks, soils, and water on a scale that far exceeds all natural processes. Developments in the technologies of energy, agriculture, health, and water supply have enabled us to multiply on an incredible scale. As human population grows by almost ninety million a year, and moves towards ten billion in the next century, humanity's influences will have an even more profound effect.

Only recently have we really become aware of what we are doing—you cannot hide from a satellite! I often call the present situation the per-cent per-year syndrome. These are among the phenomena we, as observers of our own system, now see:

• We are changing the composition of the atmosphere: carbon dioxide is increasing owing to the burning of fossil carbon at 0.3% per year. The global temperature is therefore rising, and the sea level is also rising.

• Erosion of topsoil is occurring at 0.7% per year, and this rate will continue to increase as we continue to cut down the remaining forests of the world. In Africa, about thirty trees are cut for every one planted annually. Eventually, by decreasing crop production, this can lower nutrient supply and lead to massive starvation.

• We are destroying the existing living species—an estimated twenty per cent of them in the next decade. This is perhaps most tragic of all, because it is the great array and diversity of species that provides the stability of our environment systems.

We could add a host of other items to this list, but the question we must ask is: Can we survive this onslaught on nature?

We look at our past history for some clues. Micro-organisms like bacteria flourished alone for 3.2 billion years of Earth's history before more complex life developed. There are many possible reasons why that evolution was so slow. The early oceans may have been too hot. The turbulent volcanism may have caused climate variation. Ultraviolet radiation from the early Sun may have been too intense. Of one thing we are certain: complex life is fragile—we have been again reminded of this by the AIDS epidemic of the 1980s.

Can we survive our own onslaught on the nature and character of our planet? A climatic aberration such as a year without a summer—a little ice age, with a population of ten billion people—could be catastrophic. So would be a five-per-cent warming in global temperature, bringing among other calamities sea-level rises that would innundate the homes, farm lands, and transport facilites of millions of people. We don't really know how much flexibility life on Earth has in adjusting to global change, although we do know that it would be reduced by the continued elimination of different species and therefore of the numbers that can adapt to change.

We must ask if we can find within ourselves the ability to change, to see ourselves as part of the whole, and to learn how to harmonize within it, to manage ourselves so we don't destroy our own species and our planet.

# Gaia

James Lovelock, an independent British scientist, developed the Gaia hypothesis in the late 1960s. His theory is that the Earth is a self-regulating system able to maintain the climate, atmosphere, soil, and ocean composition in a stable balance favourable to life. He named this process after the Greek goddess Earth.

A Fellow of the Royal Society of London, Lovelock is an

inventor and chemist with a doctorate in medicine who developed an electron capture detector. This led to the discovery of pesticides in penguin fat and breast milk in the early 1950s, and provided data for Rachel Carson's *Silent Spring* (1962), which was one of the first books to protest the use of weed-killers and insecticides as a hazard to both wildlife and humans.

Lovelock and his colleagues have now brought the Gaia hypothesis to the point where it is seriously considered in the scientific community. Moreover, some religious people, including Christians, have embraced his ideas as an approach to stewardship of the Earth.

The Gaia idea came out of research Lovelock was doing for the US National Aeronautics and Space Administration (NASA) to see if there was life on Mars. Using his invention, the telebioscope, he saw that there was not. When he applied the same method to the Earth, he saw a mixture of gases that could be maintained only by continual input from biological processes—in other words by life, which in a natural cycle keeps the atmosphere, oceans, and land surface fit for life.

This contrasts with more conventional scientific thought that as planetary conditions change, life evolves and adapts to fit them, or dies. By Gaia principles it is not precisely an inability to adapt that might kill a species, but a failure to play along with the rules. Gaia, suggests Lovelock, would be ruthless in eliminating species that did not contribute to keeping the system fit for life.

The Gaia hypothesis has helped explain some geophysical phenomena that affect life. The Earth's surface temperature, for example, has remained relatively constant over the four billion years since life first appeared, despite the fact that the Sun's heat has increased twenty-five per cent. (The atmospheric carbon dioxide level dropped over the same period, reducing the heat-holding greenhouse effect.) Atmospheric oxygen has been held at its current twenty-one per cent for at least 200 million years by the complex interactions of plants, animals, and elements. The salt content of the oceans also remains constant, despite the amount of the mineral that washes continually from the land.

If Gaia is modelled in sufficient detail, will we be able to predict the impacts of our actions? 'There can be no prescription, no set of rules for living within Gaia,' Lovelock writes. 'For each of our different reactions, there are only consequences.'

How does Gaia tolerate humankind's incredible influence on the face of the Earth and its elemental cycles? What kind of relationship will we have in the future? Lovelock offers two options. First, many of us are well aware of the dangers of global pollution of the oceans and atmosphere, and we have sophisticated monitoring technology to record Earth's health. We can discover what we are doing and respond accordingly. Alternatively, our disturbances of the Earth's balance will cause a rapid change in global climate and composition, with a dramatic change in the evolutionary process, as when the dinosaurs vanished. New flora and fauna will appear, fit for the Earth's new stable state. It may not include humans.

The tropics, the continental shores of the oceans and seas, and wetlands are critical contributors to Gaia's life-sustaining properties. These are the very areas at risk under the deforestation and greenhouse-warming predicted by many scientists today—we neglect to care for these regions at our peril, and Gaia's.

If Gaia is alive, continually responding and reacting to feedback like a biological system, does she have a consciousness? Does she have intentions?

Lovelock calls the thought and emotion elicited in our relationship with the planet 'the most speculative and intangible aspects of the Gaia hypothesis'. He seems to back away from implications that Gaia involves foresight. 'It's just a kind of automatic process,' he says.

BEV McBRIDE

CHAPTER 2

# The Changing Atmosphere

## GORDON A. McKAY and HENRY HENGEVELD

When the first men who circumnavigated the moon in 1968 saw the planet Earth from that vantage-point, they were awed by its beauty and vulnerability. The atmosphere seemed to be so thin as to be almost invisible. Nevertheless that same atmosphere nourishes and protects life on Earth. This is accomplished through:

• an abundance of oxygen, which constitutes twenty-one per cent of Earth's make-up and is vital to life as we know it;

• a protective layer of ozone, shielding us from the damaging effects of the sun's ultraviolet rays;

• concentrations of carbon dioxide and water vapour that are well suited for photosynthesis, the process by which plants use sunlight to produce food and fibre;

• maintaining a 'greenhouse effect', keeping Earth's surface temperatures at the right warmth so that water remains liquid.

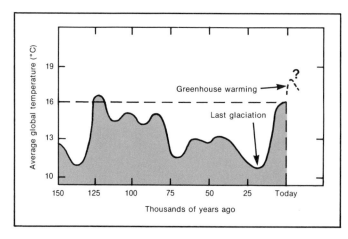

FIGURE 3. *Historic climate change. A change of only 6 degrees C in global temperature separates ice ages from warmer periods. Anticipated global warming over the next fifty years could exceed anything experienced in human history. (Modified from J.M. Mitchell Jr., 1977.)*

This balance, over the past several hundred-thousand years, has consistently remained within limits that help life thrive. Climates have varied in that time, from ice ages to warm periods like the present, but always the mean-surface temperature of the whole planet has stayed within a range of only seven degrees Celsius, from the coldest to the warmest times (Figure 3). Concentrations of carbon dioxide—that vital elixir on which so many life processes depend—have varied from 200 parts per million in glacial periods to 300 in interglacial times. Methane, another important greenhouse gas, has also remained relatively constant over the last 10,000 years. Through natural processes, the atmosphere has always maintained the delicate balance necessary for life on the planet.

## THE THREAT OF INADVERTENCE

That humans can pollute the air they breathe is not news. With industrialization, the quality of air over cities has degraded disastrously, becoming more and more toxic. The December 1952 smog in London, England, killed 4,000 people, mostly the old and the very young, and doubled the frequency of respiratory diseases. Smelters have laid waste vegetation over large areas such as the Columbia River Valley and around Sudbury, Ontario. Air pollutants have damaged health, property, vegetation, and have offended our aesthetic senses.

Until recently these ills were relatively local, as were the remedies. The 1952 London smog, for example, was a watershed event that led to the British Clean Air Act, later copied in other countries. One solution often adopted, then and now, is to time offensive emissions to coincide with weather that would ensure their escape from the area. Another is to build high exhaust stacks so that the effluent could be diluted in the atmosphere before it reached ground-levels well downwind. But solving local problems by using the atmosphere as an infinite sewer has only led to regional problems. Acid raid is an example. But among industrial effluents there are other gases, the so-called greenhouse gases, which we have assumed to be inoffensive because they are invisible and odourless. They include carbon dioxide, chlorofluorocarbons (CFCs), and nitrous oxides, which among other gases rapidly mix with the air and are assimilated into the global atmosphere. We have lived with most of these chemicals since life began, but not in the enormous quantities, or the mixes, we have now created.

It is all too clear that the global atmosphere is not a limitless ocean of gases into which we can endlessly spew our refuse without experiencing any consequences. The chemical balance that is so critical to life is being changed at an alarming rate, and the statistics are jolting

*Heating with fossil fuels adds to the pollution and contributes greenhouse gases to the atmosphere. (SSC/Photo Centre/ASC.)*

*Sulphur-dioxide pollution from smelters has laid waste vegetation over large areas—including, in the past, around Sudbury, Ontario. (SSC/Photo Centre/ASC.)*

us out of our complacency. Almost unknowingly, we have increased the atmospheric carbon to a level unprecedented over at least the past 100,000 years, with a ten per cent rise in the past thirty years alone. We have done this mostly by cutting down forests—trees absorb carbon dioxide and give out oxygen—and tilling agricultural lands, and by burning fossil fuels, which emit carbon dioxide and accelerate the warming effect of Earth. At the same time, methane and nitrous oxide, both naturally occurring greenhouses gases, are increasing annually by one per cent and 0.3 per cent respectively. Man-made CFCs, which affect the ozone layer as well as the Earth's climate, are increasing at a rate of five to six per cent a year. Our protective ozone shield in the upper atmosphere is thinning, particularly in polar regions. Acid rain and airborne toxic chemicals are increasing globally. The Arctic, once pristine, is now blanketed in winter with a haze of pollutants mainly from industrialized Eurasia. Air-borne pollutants now cross oceans, reaching as far as Antarctica.

The atmospheric 'sewer' happens to be part of the global commons within which we live and breathe, a vital support system moving across and shared by all parts of the Earth. Our assault upon the purity of air is an assault on our own health. Not only are the atmospheric changes global and significant, they are in many respects irreversible. They can radically change our environment, particularly our climate. How the global system will respond we are not sure, so the precise results are still largely unpredictable. But the fact is that the changes are uncontrolled, unplanned, unintended—and the results are potentially enormous.

## CLIMATE CHANGE: ITS CAUSES AND EFFECTS

Climate interacts intimately and systematically with all other components of the natural environment. It influences the character of land and ocean surfaces and in turn is shaped by them. Figure 4 shows the

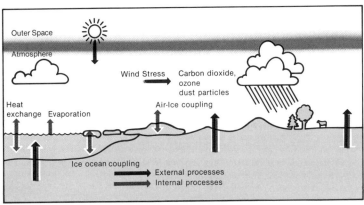

FIGURE 4. *The climate system, the interaction of five physical components: air, water, ice, land, and vegetation.*

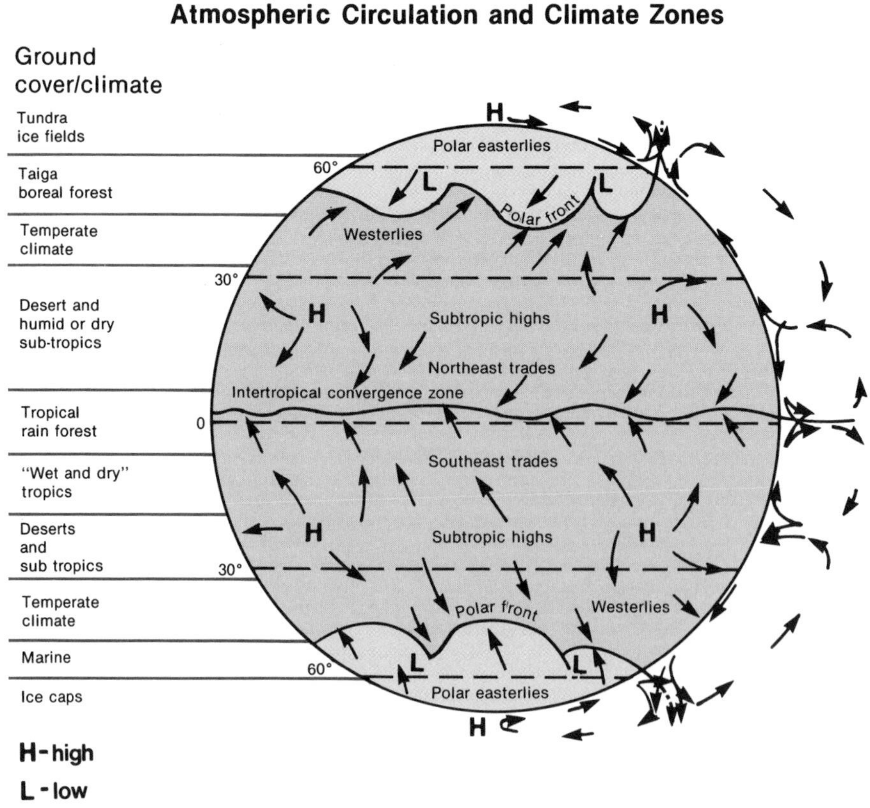

## Atmospheric Circulation and Climate Zones

Ground cover/climate

| Ground cover/climate | |
|---|---|
| Tundra ice fields | |
| Taiga boreal forest | |
| Temperate climate | |
| Desert and humid or dry sub-tropics | |
| Tropical rain forest | |
| "Wet and dry" tropics | |
| Deserts and sub tropics | |
| Temperate climate | |
| Marine | |
| Ice caps | |

- 60° — Polar easterlies — H
- L — Polar front — L
- Westerlies
- 30°
- H — Subtropic highs — H
- Northeast trades
- Intertropical convergence zone
- 0
- Southeast trades
- H — Subtropic highs — H
- 30°
- Polar front — Westerlies
- 60° — L — L
- Polar easterlies — H

**H**-high

**L**-low

FIGURE 5. *Atmospheric circulation (omitting the differential heating of land and oceans and the influence of mountain ranges). Arrows indicate generalized vertical movement toward the poles and the equator. The left-hand column shows related climate and notable vegetation or land surface. (Adapted from H.J. Critchfield,* General Climatology, *Fourth Edition, Prentice Hall Inc., 1983.)*

main components. Since everything within the system interacts, we can see that changes in land-use, atmospheric chemistry, and the ability of the atmosphere to absorb and emit heat all modify climate itself. Alter one component, and the others will adjust to a new equilibrium—and alter the global atmospheric circulation that shapes regional climates (Figure 5). That is why subtle climate variations can induce calamitous changes in ecosystems, economies, and the fortunes of nations.

To understand what is happening, we need to grasp some basics about how the atmosphere and climate systems behave, and how human activities are altering them. That alteration is evident in three ways that have potentially enormous implications for society: climate warming, stratospheric ozone depletion, and toxic and acid depositions.

## CLIMATE WARMING

Traditionally we have defined climate by the long-term statistics of variables, like temperature and humidity, used to describe the weather in a geographic area. Perhaps that's why we tend to think of climate as a constant, for changes over a lifetime are often barely perceptible. In reality, climate is far from constant. It is always changing, because of its astronomical controls and their interactions, and also because of natural internal changes in the atmosphere, oceans, and land-surfaces that cause abnormal regional weather. The El Niño (see pages 71-2) is such an irregularity. Understanding these dynamics, and being able to predict climate—including how it is influenced by human activities—require a systems approach. Because of the large number of influencing factors and complex interactions, we fall back on numerical predictive models.

The run of weather over a series of years, used to define climate, includes a wide range of extremes and unusual episodes. The media frequently report the disastrous effects, such as regional famines, desertification, forest fires and floods, heat waves and cold snaps, droughts, and pollution episodes. These are not necessarily signs of a changing climate, but they can vividly foretell what might lie ahead if the climate were to shift drastically. There may be some advantages to change, but sudden cataclysmic change usually proves extremely stressful and costly to most people. This is why we are so passionately concerned about climate today: we need to avoid, or at least relieve, the effect of change resulting from our own actions.

Climate changes have always been with us, but during the last three million years they seem to have been relatively muted, with average global temperature swings of only 7°C or less. When most of Canada was covered by ice 10,000 years ago, the air temperature of the whole planet averaged only 5°C less than today. Shifts in the amount of energy from the sun received and/or absorbed at the Earth's surface are generally involved in these long-term temperature changes. Alterations may be caused by cyclical changes in the wobbling of the Earth's axis, its tilt, and the shape of its orbit around the Sun—all with very long cycles, up to a hundred thousand years.

More detailed analyses of data from the past 20,000 years suggest that shorter, unexplained fluctuations also have happened—resulting, for example, in events such as the Little Ice Age of AD 1400 to 1850. Natural catastrophes can also cause short-duration fluctuations. For example, in 1816 the northern hemisphere experienced a 'year without a summer', probably caused by volcanic emissions from the Tambora volcano in Indonesia. Gases and fine dust from Mexico's El Chichon eruption in 1982 reduced by up to five per cent the intensity

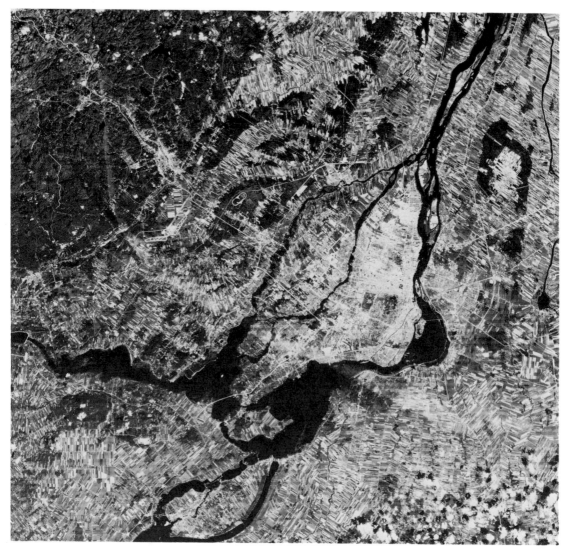

*Cities create their own geography and their own micro climate, as shown by this LANDSAT image of the Montreal region. In the upper left can be seen the transition from the Canadian Shield to the St Lawrence lowlands. The white band across the image is the vapour trail of an aeroplane. (Courtesy Canada Centre for Remote Sensing; Surveys, Mapping and Remote Sensing Sector, EMR Canada.)*

of the sunlight over a wide swath in the northern hemisphere. But such a darkening of the sun is short-lived when caused by a single volcanic eruption, since the atmosphere is virtually cleansed within two years.

We have always accepted changes in climate and its fluctuations as something that, like death and taxes, are inevitable, natural, and to be tolerated. Now we are being forced to look at the idea that human influences can produce changes in the natural climate system that may be greater and faster than anything we have yet seen. Industrialization and growth can significantly alter the energy exchanges of the atmosphere by altering its chemistry. Not only may they threaten our climate, but they could alter the productive capacity of the earth and the diversity and health of life, including human health (Figure 6).

## THE ENHANCED GREENHOUSE EFFECT

Within the atmosphere are small amounts of a number of important gases, popularly called greenhouse gases, because they alter the flow of life- and heat-energy through the atmosphere, much as does the glass shell of a greenhouse. Their effect on incoming solar energy is minimal, but collectively they act as an insulating blanket around the planet. They do this by absorbing and returning to the Earth's surface much of its outgoing heat, trapping it within the lower atmosphere. A greenhouse effect is natural and essential to a livable climate on Earth. Without it, the blanket would become a shroud. Earth's surface would be more than 30 degrees colder, the difference between a planetary climate well suited to life and a cold, lifeless sphere like the moon.

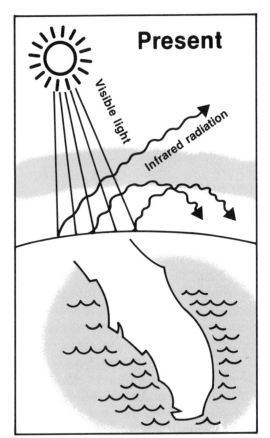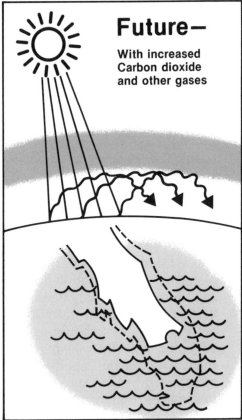

FIGURE 6. *The greenhouse effect. (Atmospheric Environment Service of Canada.)*

Greenhouses gas concentrations, however, are being drastically affected by human activities. One of the most important gases, carbon dioxide, is an important nutrient for plants, but it is potentially dangerous to our climate if enormously augmented. Its concentration has increased from about 280 parts per million in 1850 to about 350 today, mainly by greatly increased burning of fossil fuels, forest removal, and agriculture (Figure 7). (Massive forest removal in Brazil has made that country the world's third largest emitter of greenhouse gases.) Other gases, such as nitrous oxide (produced by fossil fuels and chemical fertilizers), methane, and surface ozone, although they are less abundant, are also increasing rapidly and are potentially dangerous. Man-made chlorofluorocarbons (CFCs) are used, among other things, as coolants in refrigerators and air conditioners, for making foam plastics, and as cleaning solvents for microelectronic circuitry. The most common industrially produced CFCs, although measured in parts per trillion, are among the most potent and the most rapidly increasing greenhouse gases in existence. One free chlorine atom, produced in the stratosphere by the effects of ultraviolet light on CFCs, can eliminate 100,000 molecules of ozone!

The principal causes for the increases in greenhouse gases include the rapid rise in human and domestic animal populations; greater burning of coal, oil, and natural gas for energy; deforestation; modern agricultural practices; altered land use; and industry (Figure 8).

The result? Increased concentration of greenhouse gases enhance the global greenhouse effect, trapping more heat near the Earth's surface. A warmer atmosphere can hold more water vapour, itself a powerful greenhouse gas, amplifying the warming. (On the other hand, the increase in airborne moisture may mean more clouds, which could cut off sunlight and limit or modulate the warming.)

On the basis of climate models, some scientists predict a potential increase in global surface temperature of between 1.5 and 4.5°C in the next fifty years. This may not seem like much, but 4.5°C equals the total temperature rise since the peak of the last ice age 18,000 years ago, and the increase will be even higher in some regions. The average could be slightly lower in the tropics, but at least doubled at high latitudes—mainly because of the disappearance of ice and snow. Snow reflects back as much as 85 per cent of the Sun's rays. Snow-free land-surfaces absorb more of the Sun's rays than snow surfaces, reflecting only 8 to 40 per cent, so that warming by the Sun will increase as the duration of snow-cover diminishes. Moreover, the increases in temperature can be quite large where there is relatively low energy from the Sun because of the very shallow, strong temperature inversions typical of the Arctic cold season. Reduced ice-cover on

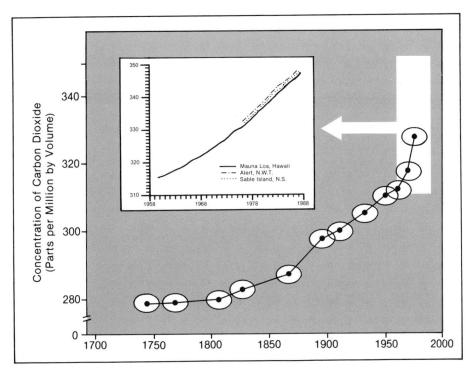

FIGURE 7. *The concentration of carbon dioxide in the atmosphere from 1740 to the present. (Atmospheric Environment Service of Canada.)*

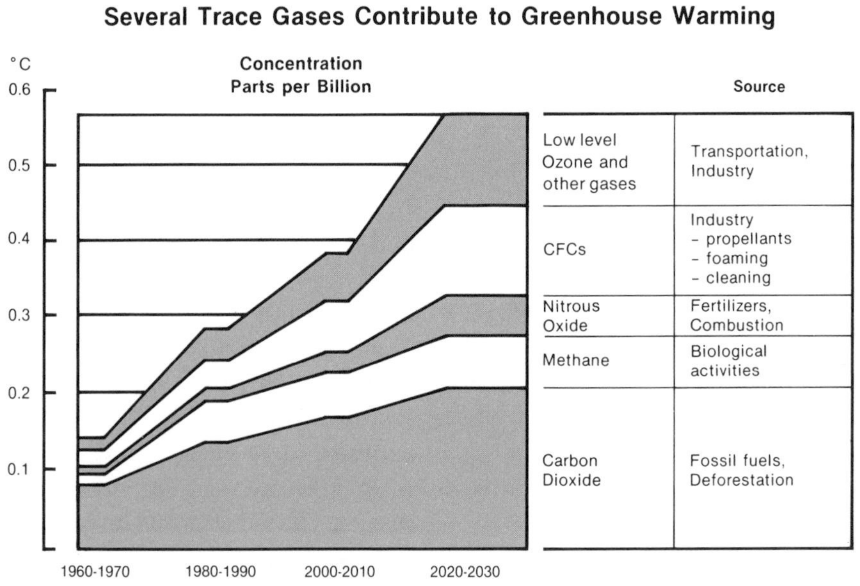

## Several Trace Gases Contribute to Greenhouse Warming

| °C | Concentration Parts per Billion | | Source |
|---|---|---|---|
| | Low level Ozone and other gases | | Transportation, Industry |
| | CFCs | | Industry<br>– propellants<br>– foaming<br>– cleaning |
| | Nitrous Oxide | | Fertilizers, Combustion |
| | Methane | | Biological activities |
| | Carbon Dioxide | | Fossil fuels, Deforestation |

1960-1970  1980-1990  2000-2010  2020-2030

FIGURE 8. *Some trace gases contributing to greenhouse warming. (World Resources Institute.)*

the polar seas will also increase the heat transfer from water to the overlying air.

Great shifts in global circulation and precipitation patterns would accompany higher temperatures, but we are still uncertain about the nature of the changes and their effect on specific regions. The global mean temperature has increased about half a degree over the past century, and the sequence of warm years in the 1980s includes the five warmest years on record. This trend is in phase with the increase in greenhouse gases; but because climate varies so much naturally, the cause is not proven beyond a doubt. Confirmation will probably take another ten years.

Predicted Number of Annual Days When the Temperature
Will Exceed 40°C (100°F) for
SELECTED NORTH AMERICAN CITIES BY 2050
(Parenthetical figures indicate the number of days
in which 40°C (100°F) is now exceeded)

| | | |
|---|---|---|
| Washington D.C. | 12 | (1) |
| Omaha | 21 | (3) |
| New York | 4 | (0) |
| Chicago | 6 | (0) |
| Denver | 16 | (0) |
| Los Angeles | 4 | (1) |
| Memphis | 42 | (4) |
| Dallas | 78 | (19) |
| Winnipeg | .1 | (0) |

Humankind can also change climate on a local scale in a variety of other ways. We have inadvertently moulded the Earth's climate through the centuries by forest removal, agriculture, urbanization, water-diversion systems, irrigation, and other large-scale activities. Any change in surface cover or use that alters the exchange of moisture and heat between land and air can change the local climate. This may be indirect, as when we reduce stream-flow into an estuary by upstream water control or diversion. Or it may be direct, as when we replace vegetation with cities. Paved roads and parking-lots absorb much more heat from the Sun, and become much hotter than forests or grasslands, which reflect more sunlight and are also cooled by the exhalation of vapour from leaves. When we add the direct escape of heat from buildings and cars we get heat islands, which can merge to alter the regional climate of highly urbanized areas, such as the Boston-Washington megalopolis. The average temperature of both Toronto and Paris, for instance, has risen about 1.5°C over the past century because of urbanization, and in most cities it is not rare, on a still night, for temperatures to be 10°C higher than in the surrounding countryside. Also, as cities get larger, they and their environs can get worse and more frequent thunderstorms, and change in both wind regimes and rainfall patterns.

## STRATOSPHERIC OZONE DEPLETION

Stratospheric ozone, a form of oxygen, occurs naturally in the cool upper atmosphere and forms an effective sunscreen by absorbing much of the ultraviolet radiation coming from the sun. Certain gases released into the atmosphere attack ozone, decreasing its concentration and thereby its ability to shield us and the Earth's ecosystems from these harmful rays (Figure 9). These gases are also major contributors to the greenhouse effect.

## Protective Ozone Layer

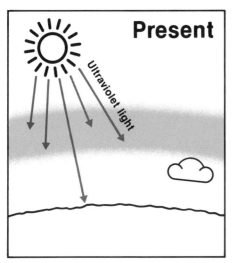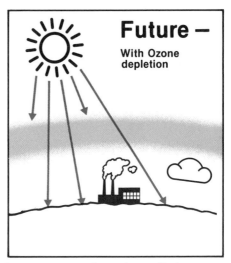

FIGURE 9. *An ozone layer in the high atmosphere screens out damaging radiation.* *(Atmospheric Environment Service of Canada.)*

The threat to this fragile shield by human activities was dramatized in 1985 when British Antarctic Survey scientists made the startling observation that concentrations of ozone over Antarctica were dropping at an astounding rate each spring. Since then a marked reduction in the total ozone has been observed over Antarctica each spring. This ozone 'hole' is about half the size of Canada. (Figure 10.)

Back-tracking the satellite records reveals that there has been a fifty per cent loss of ozone over the ten years from 1977 to 1987. A smaller hole, about one-third the size, has been observed over the Arctic. But the reduction is not confined to the poles; the ozone layer seems to have been slowly depleting around the globe, especially in temperate and subpolar zones, with reductions of two or three per cent. Diminishing the ozone layer over populated areas could have dire consequences. Ultraviolet radiation is invisible, but it causes sunburn and skin cancer, and has been linked to cataracts, a major cause of blindness, and—most dangerous—a weakening of the immune system. Each one per cent drop in ozone concentration is estimated to result in an increase in skin cancers of about four to six per cent. The reduction of ozone also damages vegetation and ecosystems, particularly in water.

What is the cause of this depletion? Mounting evidence points to chemical changes in the atmosphere. We have already identified methane, nitrous oxides, and CFCs as greenhouse gases. Along with halons (man-made organic compounds used as fire retardants), they are precursors to free ions and oxides that can catalyze the destruction of stratospheric ozone. All are increasing in atmospheric concentration. Some of these gases live for hundred of years, and on reaching the upper atmosphere will continue to wreak damage for generations to come. There are other influencing factors. The role of stratospheric ice-crystal clouds over Antarctica is being studied, as are changes in stratospheric chemical processes at very low temperatures, and the effects of the onset of polar sunlight. But the central role of the gases is undeniable.

**The Ozone Hole**

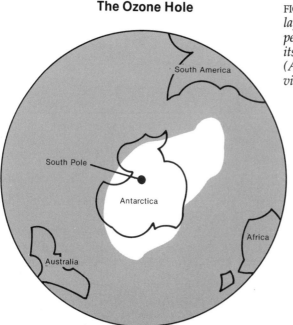

FIGURE 10. *The hole in the ozone layer over Antarctica as it appeared in 1986. The hole reached its largest size in October 1987. (Atmospheric Environment Service of Canada.)*

## TOXIC AND ACID DEPOSITION

The atmosphere is a huge storehouse for gases and particles, but it also acts as a conduit for pollutants. These are ultimately deposited as acid rain, snow, or dry matter on land and water, acidifying lakes and attacking forests. For instance, the atmosphere transports one quarter of the nitrogen entering Chesapeake Bay from the heavily industrialized Baltimore-Washington area. And a study of Lake Superior

### SOURCES AND IMPACTS OF IMPORTANT TOXIC AND ACIDIC AIR POLLUTANTS

| POLLUTANT | IMPACT | SOURCE |
|---|---|---|
| Acid Deposition (Acid Rain) | Threatens public health. Acidifies streams, lakes and soils. Damages buildings and materials. Together with ozone is implicated in death of trees. | Sulphur and nitrogen oxides from fossil fuel combustion—mainly in power plants and vehicles. |
| Surface Ozone | Principal component of smog and important greenhouse gas. Threatens public health. Injures trees, crops, and other vegetation. | Forms from atmospheric reactions between nitrogen oxides and organic compounds. |
| Organic Compounds | Help ground-level ozone form. | Vehicles, fuel-burning, industry. |
| Carbon Monoxide | Threatens public health and helps ozone form. Indirectly contributes to greenhouse warming. | Mostly from vehicles. |

SOURCE: World Resources Institute

disclosed that in this less-industrialized region the atmospheric transport of pollutants and toxic chemicals to the lake was five to seven times that transported by other means—principally by water. The re-emission from water and land surfaces enables the transport of pollutants still further from their origin. Some cancer-causing polychlorinated biphenyls (PCBs) observed in Antarctica have been attributed to this secondary release. Climate—winds, stability, etc.—affects their distribution. The toxicity of the pollutants may also be altered by the chemical reactions within the atmosphere during transport, some products becoming more, and others less, poisonous.

Rain and snow cleanse the air most effectively but deposit large amounts of pollutants on land and water. Droplets about 2 millimetres across are the best collectors, so even a very light rainfall or drizzle can bring down large quantities of impurities. When snow-cover melts, chemicals deposited in it are quickly leached out, producing a phenomenon called acid shock—a very rapid rise in the acidity of run-off, reaching concentrations that can kill some aquatic life. Dry deposition is much slower, but can be significant. The dissolution of deposited particles by rain can result in chemical concentrations damaging to plants.

During polar winters there is little purging by precipitation; moreover, strong atmospheric inversions trap pollutants in a thin layer of air near the ground. These factors are blamed for the layer of haze that blankets the Arctic throughout the winter, composed of pollutants such as sulphates and heavy metals mainly from industrialized areas far to the south, as well as particles from the land and ocean. The haze is twenty to forty times more dense in winter than in summer. Although it has an acidity about a tenth of that in industrialized areas further south, it is a serious problem for Arctic ecosystems. (Figure 11.)

## ENVIRONMENTAL EFFECTS OF CLIMATE CHANGE

The face of the planet—its natural ecosystems—is likely to be profoundly affected by atmospheric change. When to the hazards to human health are added damage to agriculture, forests, lakes, and fisheries, the toll taken by pollutants is staggering. However, the risks of climate change are the most serious. Climate controls plant and animal growth, which is sensitive to heat, water, and light. It affects the seasonality and thickness of ice-cover on water, snow-cover on land, and permafrost within the land. It influences the levels of lakes and oceans, and hence water-drainage patterns and coastlines.

Climate and ecosystems are linked in many ways. The most obvious is through the effect of air temperature:
• The production of biomass—the total quantity or weight of organ-

## Major pathways of industrial pollution

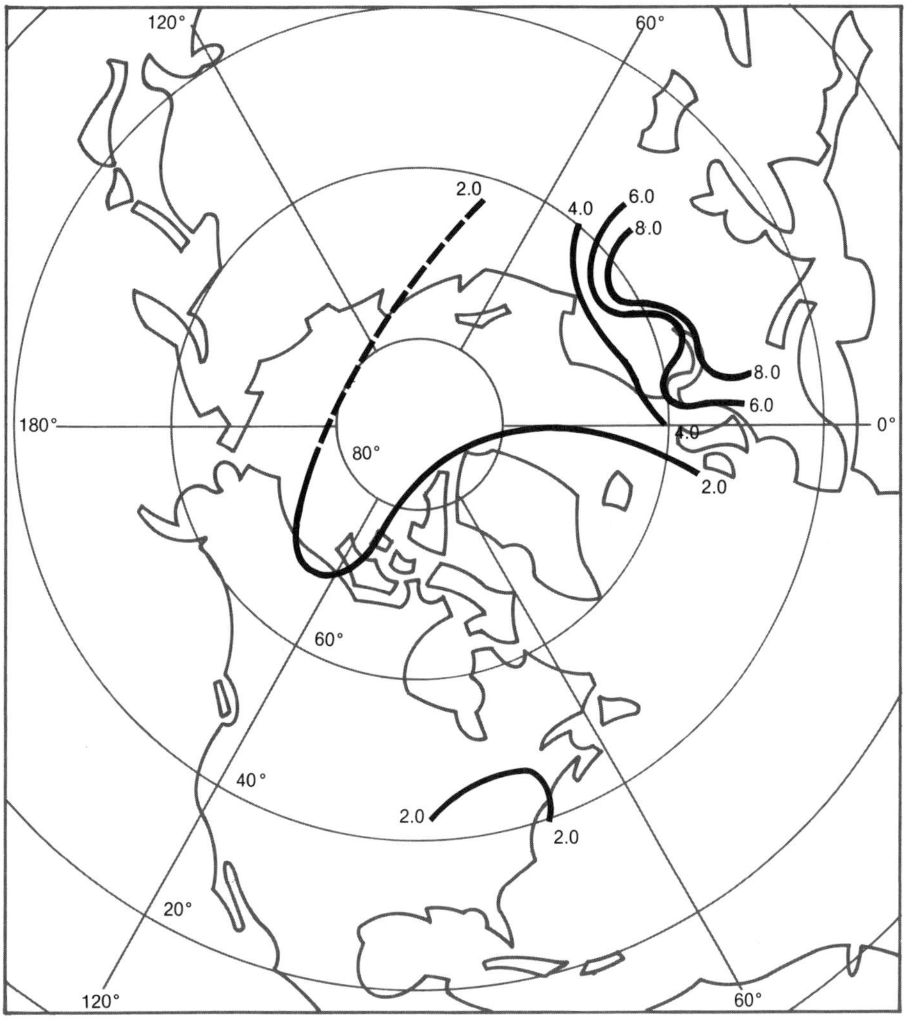

FIGURE 11. *Isolines of industrial pollution indicating sulphate content in milligrams per square metre. (Atmospheric Environment Service of Canada.)*

isms in a given area—approximately doubles for every ten-degree increase in average air temperature (assuming adequate moisture).

• Both the poleward limit of agricultural production and the edge of the sea-ice tend to fluctuate with global temperature.

• The southern limits of both continuous and discontinuous permafrost also tend to follow temperatures under the present steady climate conditions. Decay of the permafrost would follow atmospheric warming. (Figure 12.)

• The start and end of growing seasons, and of freeze-up and break-

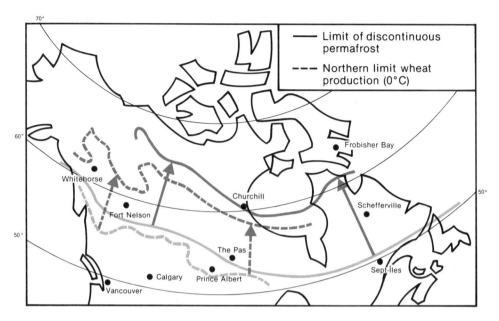

FIGURE 12. *Projected movement in the boundaries of discontinuous permafrost and the northern limit of wheat production if carbon dioxide concentration was maintained at double the present level. (Atmospheric Environment Service of Canada.)*

up dates of ice in bays and lakes, are related to the annual temperature regime.

• The warmer climate would warm the ocean, causing it to expand and increase its mass as land ice melts.

The consequences of climate change could be out of all proportion to the causes. Both geological and recent history tell us that very subtle shifts in climate can bring about major environmental changes, and these past episodes can teach us something about defending ourselves against climate-related hazards. In centuries past, global water resources, ice-cover, and vegetation patterns altered drastically as the result of shifts of about one degree in global air temperature. The warming in the first third of this century resulted in major events, such as the opening of the Eastern Passage to the north of the Soviet Union, development of the cod fishery off Western Greenland, and the 'dirty-thirties' drought on the North American Great Plains.

Droughts, frequently associated with heat-waves, have greatly reduced the volumes of surface water, concentrated water salts and pollutants, raised water temperatures, increased algae bloom, decimated wild-life populations, and drastically damaged soils. They have also stricken millions of people with famine and water shortages. The warmer and dryer climate projected for some regions as a result of

RIGHT: *Land caked and cracked by drought. (AES Canada photo.)*

*Eroded prairie soil during the 'dirty thirties', the result of bad farming practices. (Photo Saskatchewan Archives Board.)*

global warming would increase the severity of such droughts. (On the other hand, some arid regions may become more humid and fertile.)

Climate change could also make ecosystems all the more vulnerable to toxins and other stresses, including acid precipitation and ultraviolet radiation. Forests attacked by drought seem to be much more vulnerable to air pollution and large-scale die-back. For example, in 1976 intense drought and forest fires in Western Europe were followed by the quick degradation of forests because of pollution. Allergens and blights can develop explosively in certain temperature/humidity combinations. We can't predict the effects of climate change on them. We can, however, learn about climate-related hazards by studying abnormal episodes, seasonal shifts, and longer-term variations.

*London, England, has already built a flood-defence system spanning the River Thames. The most ambitious flood-prevention scheme in the world so far, the barrier consists of four main gates, each with a 61-metre span, and six smaller gates, all pivoted and supported between concrete piers. In a flood warning the complete barrier can be raised within 30 minutes. A timber carrier passes between the piers of one of the main gates. (Photo British High Commission, Ottawa.)*

If the globe warms, sea level will rise. Sea levels seem to be rising slowly now, and a further rise of as much as 1.5 metres has been projected by the middle of the next century (Figure 13). The consequence would be the catastrophic inundation of many heavily populated low-coastal regions—for instance a portion of Bangladesh, which has a population of about ten million (Figure 14). An estimated twelve to fifteen per cent of Egypt's arable land and population would be at risk. The Netherlands and coastal cities such as London, New York, and Miami would face relocation and adaptation that would cost millions and hurt many people. Scientists at a 1987 workshop in Villach, Austria, estimated that defence against—and only partial adaptation to—sea-level rise, like building dikes, could cost an estimated 30 to 300 billion dollars US. As well, coastal margins would erode and many coastal wetlands would disappear.

## Global Sea Level Change (cm)
## Global Temperature Change

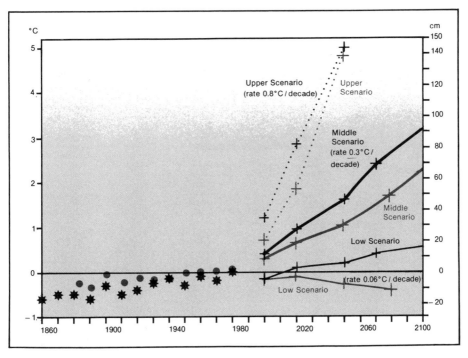

FIGURE 13. *Global temperature and sea level change, historic and projected. To 1980, based on measurements. Upper scenario: with continued growth in energy use as at present. Middle scenario: stabilized emissions to the atmosphere. Lower scenario: with drastically reduced emission of greenhouse gases. (From J. Jaeger,* Developing Policies for Responding to Climatic Change, *World Meteorological Organization, April 1988.)*

## Bangladesh sea level change

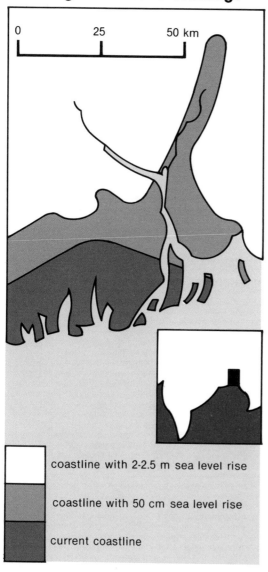

FIGURE 14. *Climate warming will cause sea levels to rise and flood many coastal areas. In Bangladesh the area that would be flooded is the most populated part of the country. UN Environment Program.*

0    25    50 km

coastline with 2-2.5 m sea level rise

coastline with 50 cm sea level rise

current coastline

The projected greenhouse warming would occur at a time in history when populations are urbanizing, especially in developing countries. We can expect this to be reflected in an increase in heat strokes and other heat-sensitive maladies, not to mention illnesses related to increased atmospheric pollution. Good air-management practices, like choosing city sites and designs to permit ventilation, and planting urban green space and trees to cool and filter the air and reduce noise, will help.

## PUBLIC-POLICY CONSIDERATIONS

The history of the development of cultures and the migration of societies clearly shows us the importance of climate, the positive influences of benign weather, and the catastrophic effects of climatically hostile periods. The recent famines in the African Sahel, the Grecian heatwave of 1987, and the widespread influence of El Niño (see page 71) in 1982-3 show us that these influences are still powerful today.

Extreme or unusual events like El Niño will probably be even more important in the future. Although their past occurrences can provide some insight into the dynamics of the atmosphere, history may not be the best guide in planning for future change. The changes predicted will be different from any we have dealt with in the past, and their rates more than we have ever experienced. If the rates of ecological change accelerate, so will the rates of change of a whole host of other factors like population, health, economic investment, and technologies. A hundred years could force upon us changes greater than have been accommodated in tens of thousands of years past. Can people and the biosphere cope with such extreme and possibly rapid change?

Atmospheric change must now be reckoned with as a major force in the future of the planet. Many scientists believe that the warming process has already begun and will continue as long as the greenhouse gases increase. Stratospheric ozone depletion is in progress, and toxic chemicals and acid depositions exact a serious toll on ecosystems, health, and materials. We will respond better if we can answer some important questions. What are the probable specific consequences? What policies could avoid or significantly mitigate inadvertent disasters? What additional knowledge do we need to answer these questions? Finding the best course for people to manage climate change may be the most vital challenge of this and the next century.

If climate change and its consequences could be predicted precisely, we could put into place logical and timely action plans. However, climate prediction is far from exact—although it is generally better than economic forecasting, which many of us are willing to bet on! We can foretell the long-term effects of an altered atmosphere in a coarse and macro-scale manner, other major forces aside, but we cannot foresee the magnitude of altered regional characteristics and year-to-year variations. Long-range weather-system movements may never be predictable. Certainly we must resolve fundamental gaps in our understanding of the climate system and its interaction within our dynamic geosphere and biosphere before we can greatly improve our abilities to prophesy.

The other great area of ignorance is the human dimension of global change, whether in social response, economics, or cultural vulnerability. Co-operative international efforts such as the International Geosphere-Biosphere Program, Human Dimensions of Global Change, and the World Climate Program are beginning to look at these concerns, but their success depends on similar national initiatives.

## POLICIES AND ACTIONS

These uncertainties—of climate prediction and the consequences of change—will not diminish within a decade, and they are often given as reasons for policy inaction, but this reaction could be suicidal. We cannot afford to wait even a decade to plumb them. The problems could become insurmountable even before we start to remedy them.

Speaking about different scenarios for greenhouse warming at the Toronto conference on the Changing Atmosphere in 1988, meteorologist Howard Ferguson, of Canada's Atmospheric Environment Service, said: 'Those who advocate a program consisting only of additional research are missing the boat.' He made the point that we must not only plan for greenhouse and other contingencies, but take some action even as research continues. Otherwise we may irreparably jeopardize our life-support system.

That 1988 Conference showed clearly that there is no simple solution to the problem of atmospheric change. All countries are involved, but many are ill-equipped to respond because of internal socio-economic problems. The international community must identify the regions and people at greatest risk, and vigorously pursue the course of developing technological and economic relief. Developed countries, the main users of fossil fuels, can and must implement effective energy and other policies that will slow atmospheric change. Many of these policies make good economic sense—witness the fuel efficiency of automobiles today, which is double that of the mid-1970s.

# Policy/Action Recommendations

1. Develop global and national action plans for the protection of the atmosphere.

2. Initiate development now of an International Convention for the Protection of the Atmosphere—while acting on regional and bilateral conventions.

3. Tax the use of fossil fuels and use the proceeds for a World Atmosphere Fund to implement plans.

4. Ratify, implement, and strengthen the 1987 Montreal Protocol on substances that deplete the ozone layer, acting toward their virtual elimination by the year 2000.

5. Reduce $CO_2$ emissions by twenty per cent by the year 2005, by conservation and supply management (Figure 15).

6. Increase research and development directed to low and non-$CO_2$-emission energy options.

7. Reduce other atmospheric greenhouse gas emissions.

8. Vigorously apply existing technologies to reduce acidifying and toxic emissions to environmentally sustainable levels.

9. Increase research and monitoring to understand and arrest atmospheric degradation and mitigate effects of change.

10. Promote public awareness through information and education.

SOURCE: The Changing Atmosphere Conference, Toronto, June 1988

## Government subsidies of public transport

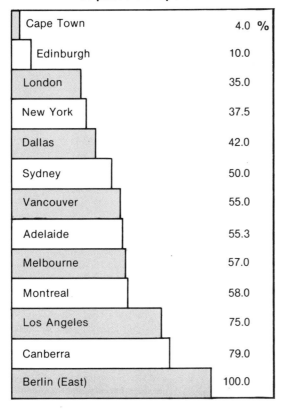

| City | Percentage |
|---|---|
| Cape Town | 4.0 % |
| Edinburgh | 10.0 |
| London | 35.0 |
| New York | 37.5 |
| Dallas | 42.0 |
| Sydney | 50.0 |
| Vancouver | 55.0 |
| Adelaide | 55.3 |
| Melbourne | 57.0 |
| Montreal | 58.0 |
| Los Angeles | 75.0 |
| Canberra | 79.0 |
| Berlin (East) | 100.0 |

FIGURE 15. *Percentage of government subsidies of public transport. Conservation and supply management can reduce $CO_2$ emissions.* (Jane's Urban Transport Systems, *1985.*)

## PERSONAL AND COMMUNITY INITIATIVES

In the end, attaining sustainable development that meets the needs of the present without compromising the lives of our children and grandchildren depends on you and me. We do influence policy-making, including policies that help us improve our understanding. Through personal and group action we exert a powerful influence on politicians, manufacturers, and our neighbours. But we are also the consumers. Burning fossil fuels, for instance, is a major factor that has got us into this predicament and is making it worse. You and I burn fossil fuels for space-heating, transportation, water-heating, refrigeration, and air-conditioning, as well as in manufacturing. Unless we can set an example in our own use of energy, we cannot expect to be credible in the eyes of others, including those less affluent than ourselves. The same applies to other actions that degrade our atmosphere. Through collective action we can contribute to sound stewardship of this planet.

# What Can I Do to Protect the Atmosphere?

*SHOPPING:*
- Choose environmentally friendly products, such as paper bags opposed to plastic bags, and items that require little energy to produce.
- Avoid 'foam' plastics made using CFCs.
- Let your buying speak to manufacturers and suppliers.

*AT HOME:*
- Recycle
- Implement energy-conserving measures in space-heating and air-conditioning, water-heating and refrigeration.

*TRANSPORTATION:*
- Opt for walking, car-pools, public transit, low-fuel consumption vehicles.
- Ensure that the vehicles you travel in conform to air-pollution standards.

*LEISURE:*
- Opt for sports that use bio-energy, not fossil fuels.

- Compost waste. Be prudent in the use of gardening fertilizers and agricultural chemicals.

*IN BUSINESS/INDUSTRY:*
- Incorporate the environment into your planning.
- Increase energy-use efficiency; investigate alternatives to fossil fuels; increase energy conservation; minimize pollutant emission.
- Ensure that the implications of your investment/purchasing conform to environmental values.

*IN THE COMMUNITY:*
- Write your politicians about your views.
- Let local businesses know your concerns.
- Promote recycling and the use of friendly products and practices.

# El Niño

Can the interplay of ocean currents off the coast of Peru be related to drought in Australia, India, and Eastern Brazil? In the winter of 1982-3 the links were dramatically illustrated during the most severe appearance in this century of El Niño. The more we learn about how climates act within one global system, the more we recognize that this type of relationship is possible.

El Niño was named after the Christ-child by Spanish-speaking fishermen over a century ago. It usually comes near Christmas. Unlike its namesake, however, modern literature does not credit the current with any blessings. 'El Niño' originally referred to an annual event when the less saline tropical current creeps southward through the Pacific Ocean along the coast of Peru. This warmer water deflects the cool, productive Humboldt current away from its customary pathway along the west coast of South America. While prevailing winds here blow toward the equator and usually perpetuate the upwelling of nutrient-rich waters from the sea bottom, during El Niño a poleward wind reverses this life-sustaining process.

Now scientists and the media tend to use the name El Niño for the occurrence of exceptionally warm waters in the tropical Pacific. More precisely, they speak of ENSO: El Niño/Southern

Oscillation. They know that El Niño is actually a manifestation of the Southern Oscillation—the massive shifting back and forth of atmospheric pressure between high- and low-pressure centres in the Pacific, first observed in 1924.

Scientists have yet to learn precisely how El Niño happens. El Niños occur generally with a frequency varying between five and ten years, and in some winters the related impacts are disastrous. Marine organisms, unable to survive the changed conditions, die and decompose, fouling both sea and air with hydrogen sulphide. These organisms include the fish population that sustains Peru's anchovy fishing industry, perhaps the most famous victim of El Niño. The catch declined from 12 million metric tons in 1970 to less than half a million tons in 1983. A decimated fish population means sea birds die because of lack of food, and the guano-for-fertilizer industry also suffers—a great disruption to the already poor Peruvians. Archeologists have suggested that El Niño has historically forced switches between agriculture and fishing as a means of livelihood for Peruvians.

In recent times the fish and bird populations have not quite recovered before the current's next strike. For example, the Peruvian coastal guano bird population of 28 million in 1956 plummeted to six million during the El Niño of 1957. The birds recovered to 17 million, only to be dashed to 4 million in 1965. Modern stresses imposed by fisheries management practices and environmental pollution make survival even more tenuous.

Far away from coastal Peru, El Niño has been blamed, rightly or wrongly, for droughts in Brazil and the Sahel, too-warm winters in Europe and Western Canada, other agricultural failures, and storms. The unusual and unexpected weather conditions of the El Niño/Southern Oscillation of 1982-3 are said to have cost $8.5 billion Canadian around the world.

ENSO is a climatic force that humans have neither caused nor have yet managed to understand thoroughly. Climatologists recognize these strong and anomalous oceanic and atmospheric circulations as clear demonstrations of linkages between events in one location and others half-a-world away.

Each El Niño is different in its intensity and personality. It also has a recognized counterpart: La Niña (the female child) cools the same waters that El Niño warms—and again it has no reliable pattern.

As we plan to deal with global climatic change in coming years, these 'children' will have to be considered.          BEV McBRIDE

# How Is Human Health Endangered?

Global change also means changing conditions for human health. We can anticipate dangers from stratospheric ozone depletion, global warming, and rising sea levels, although it is hard to estimate the precise results of these changes. We might see anything from new or more frequently occurring diseases, because of warmer climate, to stress- and nutrition-related ailments due to crop failures, as well as large-scale economic hardship and the forced relocation of people (Figure 16).

## STRATOSPHERIC OZONE DEPLETION

Most of the ultraviolet radiation on Earth comes from the Sun and is found in two main spectral parts, Ultraviolet A and B. UVA is not absorbed by the ozone layer and penetrates to the Earth. UVB is more damaging to biological processes, but fortunately is partly absorbed by the ozone layer. If the ozone layer is depleted, more UVB will get through.

We know that human skin can age prematurely if it is chronically exposed to sunlight. UVB may be one of the main causes of non-malignant skin cancer, which grows slowly on visible body parts exposed to sunlight. Its cure rate is high and mortality low, but it does cause anxiety and increased health-care costs. Researchers believe that this type of skin cancer is caused by long-term exposure to sunlight. Melanoma—a serious, malignant, skin cancer—is believed to be provoked by intermittent, intense sunburn—especially in fair-skinned, fair-haired people. These tumors do not occur on exposed skin, and we are not sure that ultraviolet radiation is the cause, but two new animal studies strongly suggest that it plays a role.

Scientists predict that for every one per cent decrease in stratospheric ozone, there will be a two per cent increase in UVB reaching Earth. This will cause a four per cent increase of non-malignant skin cancer in men, and a three per cent increase in women. Statistics for the frequency of this kind of skin cancer are not reliable since the blemishes are often treated in doctors' offices and are not reported. Some reports say that they are increasing, but because of better diagnosis and treatment, deaths are declining. The statistics on malignant skin tumors are more reliable. They have doubled from 2.3 cases per hundred thousand people in 1971 to 5.6 per hundred thousand in 1982.

# Health Effects of

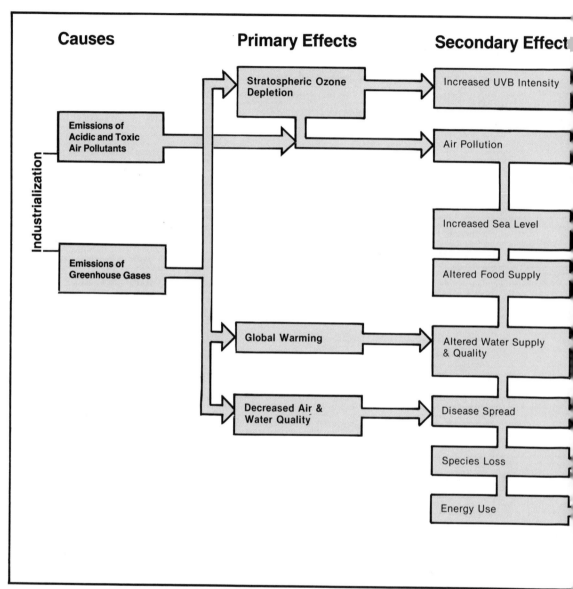

FIGURE 16.

# Global Changes

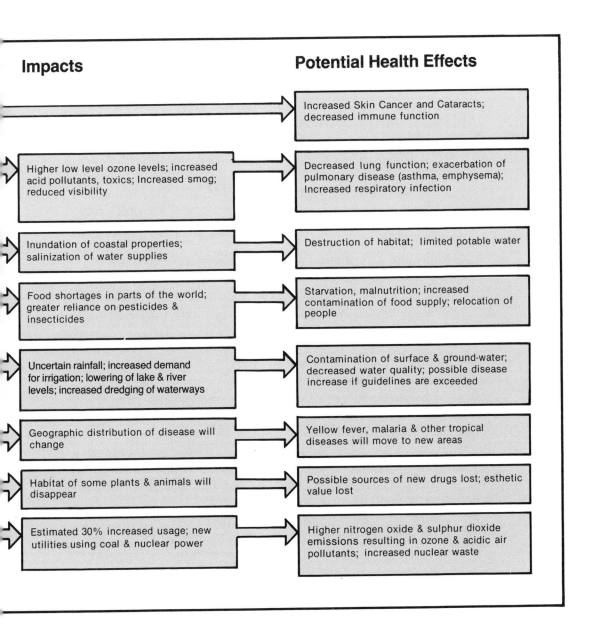

**Impacts**

**Potential Health Effects**

| Impacts | Potential Health Effects |
|---|---|
| | Increased Skin Cancer and Cataracts; decreased immune function |
| Higher low level ozone levels; increased acid pollutants, toxics; Increased smog; reduced visibility | Decreased lung function; exacerbation of pulmonary disease (asthma, emphysema); Increased respiratory infection |
| Inundation of coastal properties; salinization of water supplies | Destruction of habitat; limited potable water |
| Food shortages in parts of the world; greater reliance on pesticides & insecticides | Starvation, malnutrition; increased contamination of food supply; relocation of people |
| Uncertain rainfall; increased demand for irrigation; lowering of lake & river levels; increased dredging of waterways | Contamination of surface & ground-water; decreased water quality; possible disease increase if guidelines are exceeded |
| Geographic distribution of disease will change | Yellow fever, malaria & other tropical diseases will move to new areas |
| Habitat of some plants & animals will disappear | Possible sources of new drugs lost; esthetic value lost |
| Estimated 30% increased usage; new utilities using coal & nuclear power | Higher nitrogen oxide & sulphur dioxide emissions resulting in ozone & acidic air pollutants; increased nuclear waste |

Both the immune system and the human eye may also be affected by ultraviolet radiation. In lab experiments with animals and with human skin, UVB radiation was found to suppress the skin's immune system. We are not sure if this immuno suppression is linked to ultraviolet-induced skin cancer, but it may trigger infectious diseases like cold-sores. The cornea and the lens of the eye absorb ultraviolet radiation, and over-exposure can cause photo-conjunctivitis, a painful but reversible inflammation of the outside part of the eye. Animal experiments also show that high or steady exposure to sunlight causes cataracts. We do not know whether the levels that occur outside the laboratory are high enough to have the same effect. We also do not know what causes the majority of cataracts in older people. If these cataracts are caused by UVB and UVA, the impact could be profound, particularly in developing countries where surgical capabilities are limited.

We need more information about the effects of sunlight on skin and eyes, including better information about what amounts are dangerous. Meanwhile, reducing our exposure to UVB through preventing ozone depletion and changing our lifestyle is only sensible.

## AIR POLLUTION

The interactions between predicted global warming—the reduction of the Earth's protective shield of ozone—and increased pollution of our atmosphere are complicated, as are the effects on our health. Many people died or became ill during times of high air pollution in the 1950s and 1960s—in, for example, London, England, Los Angeles, and Belgium's Meuse Valley. Studies of these events provide the strongest evidence that human health suffers from air pollution.

The use of fossil fuel to generate electricity—to run automobiles and to produce commercial petrochemicals—releases nitrogen oxides and volatile organic compounds into the air. There the compounds are transformed through hundreds of chemical reactions into ground-level or tropospheric ozone. Ozone, so valuable in the stratosphere as a shield against ultraviolet rays, becomes a problem at ground level because it is very hazardous to human health when inhaled. Permissable ground-level ozone standards vary from country to country, and in Canada the federal government says 80 parts per billion are acceptable. The US government allows 120 parts per billion.

Monitoring and modelling show that ground-level ozone extends well beyond the local urban smog, and exposes many millions of people to unacceptable levels. Hundreds of cities in North America alone are at present not able to meet the standards set under federal legislation, especially in the summer—in Los Angeles the air quality is unacceptable more than seventy-five per cent of the time! Data from lab animals show that acute peaks cause more damage than continuous exposure. A resulting inflammation in the lungs may lead to emphysema. Studies on people show that their lungs function less well as exposure continues. This is true at the US acceptable level and lower.

Both higher temperature and greater ultraviolet-radiation intensity enhance the reactions that form low-level ozone—another link in the complexity of air pollution, stratospheric ozone depletion, and global warming.

The human respiratory tract is a vulnerable target of another set of air pollutants: sulphur dioxide and nitrogen oxides. High concentrations of sulphur dioxide are released into the air by coal burning in non-ferrous smelters and electrical facilities; nitrogen dioxide is emitted from cars. These gases turn into sulfuric and nitric acids and other products that sometimes travel great distances before they are deposited or breathed in. Victims may find breathing difficult—with coughing, wheezing, and chest pain, and, more importantly, reduced general health. Continued respiratory-tract inflammation and exposure to the chemicals may be related to chronic obstructive pulmonary diseases.

Controlled studies show lung-function changes in volunteers exposed to acidic air pollution at levels much higher than currently occur outside. Canadian school children exposed over a long period to sulfate air pollution where they live over a long period were recently found to have impaired lung function. The damage was small but statistically significant, and it is possible that it could get worse if they continued to breathe in the pollution. The Canadian Department of National Health and Welfare, and the Harvard School of Public Health, are now investigating the relationship between acid pollution and respiratory health and lung function.

We know very little about what happens when various pollutants mix together, but we have clues that they can be even more damaging in combination. Acidic pollution may make lungs more sensitive to ground-level ozone. Current levels of ozone, especially near big highways, may already have an unacceptable impact on health; any increase could be very serious.

Acid deposition affects water quality. By altering the pH and chemical composition of water, toxic metals that normally would have remained where they were may be dissolved and released into drinking water. These can include aluminum, arsenic, cadmium, and lead, from soil, ore bodies, and plumbing systems. The sources most at risk are rainwater cisterns and wells less that 15 metres deep in poorly buffered regions such as the Canadian Shield. Municipalities adjust their treated water for acidity, minimizing its effect on plumbing, but an estimated two million Canadians depend on shallow wells or rainwater, and about 300,000 of them now live in areas that may be affected by acidity. More acid deposition could increase the problem.

## SEA LEVEL AND WATER SUPPLY

If the sea level rises and precipitation patterns change as predicted with global warming, human health would again be endangered by conditions of water quantity and water quality. The impacts would be worse in areas already experiencing water problems. Salinization, groundwater shortages, release of contaminants from sludge dredged to maintain navigation channels and flooding of sanitation systems, garbage dumps, and water-treatment facilities are all predictable threats to safe drinking water as the sea level rises.

## FOOD SUPPLY

Shifts in world agricultural practices because of climate change will force people to relocate and will disrupt the present patterns of food supply and distribution. The North American population will probably be able to adapt and maintain adequate food, but if more pesticides and fertilizers are used to increase yields, both food and water could be further contaminated.

The potential health effects of such shifts, and any influx in toxics to the environment, are again difficult to predict. One problem is the time frame: cancer, for instance, has a latency period of five to thirty years; by the time it is diagnosable, it is often too late to tell what might have caused it, and too late for treatment. Genetic damage may take generations to show up, while general damage to the immune system, if not measurable, can still lower disease resistance. Modern medicine is just beginning to address the cumulative and combined effects of chemicals on human health.

## DISEASE SPREAD

Many diseases are carried by insects that need certain minimum temperatures to survive and reproduce. If the temperature increases in northern North America, we can speculate that we may experience more insect-borne diseases that we now associate with warmer climates. North Americans could expect more Rocky Mountain Spotted Fever and Lyme Disease, both borne by ticks, and more malaria, dengue fever, and encephalitis, all carried by mosquitoes. Yellow fever, hepatitis, and various tropical fevers would all likely broaden their geographical distribution into new, warmer climates, and new strains could develop.

## HEAT STRESS

Many people, particularly children and seniors, have trouble coping with extreme heat. Heat-stress-related mortality and hospitalization are likely to increase significantly with climate warming as heat spells become more frequent, more intense, and longer.

CLAIRE A. FRANKLIN

# CHAPTER 3

# Our Fragile Inheritance

## W. RICHARD PELTIER

The planetary oasis on which we live today continues to be strongly influenced by its own creation—by the way it was, immediately following its accretion out of the primitive solar nebula. Billions of years ago a star became unstable and exploded with an incredible release of energy. The 'accretion disc' of gas and dust that was probably created from this explosion was later the context for the Earth-formative events that began over 4.5 billion years ago. Within the disc-womb the dominant gases pulled together to form the Sun and to illuminate it by the hot nuclear fusion reaction in which hydrogen burns and produces helium.

As the sun formed, its gravitational field strongly guided the remaining material in the accretion disc. Most of the dust had begun to coagulate, and the swarm of objects within one zone of the accretion disc came to be dominated by planetesimals—little planets that

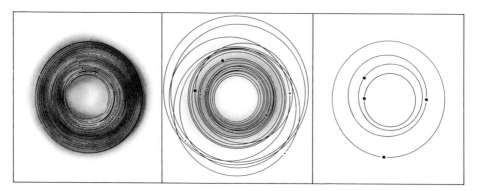

FIGURE 17. *Formation of the sun and planets. Following the explosion of an earlier star, the hot mass of the early Sun rotated around an axis, radiating heat and attracting gases and dust. This is a computer simulation of the accretion of the terrestrial planets Mercury, Venus, Earth, and Mars, beginning with a large number of planetesimal-sized objects. The process took about 100 million years—a small fraction of the Earth's age, approximately 4.6 billion years.*

we now believe must have evolved very quickly to produce the terrestrial planets themselves: Mercury, Venus, Earth, and Mars. They took up the distances they now occupy from the Sun. Further away, a residual gaseous component remained that would seed the formation of the giant planets of Jupiter and Saturn.

The best estimate of the time it took for the terrestrial planets to evolve is less than 100 million years, a small fraction of the Earth's age. Figure 17 shows a modern computer model of the way gravity brought the terrestrial planets together and formed them.

Since this final coming-together happened so relatively quickly, the Earth and other planets near it must have formed hot—in fact so hot that the initially homogeneous compositions of the individual planetesimals were completely differentiated. By the end of accretion, therefore, most of the heaviest element, iron, had sunk to the centre of the primitive planet Earth, leaving the remaining material, mostly iron-magnesium silicates, to form a thick mantle surrounding the molten iron core—the same planetary structure as exists today (Figure 18).

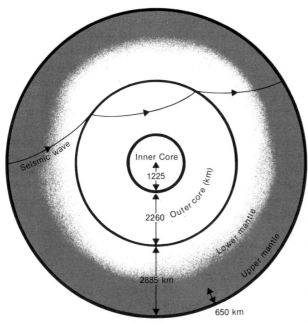

FIGURE 18. *The interior of the Earth. The curved lines traversing the structure represent the paths of travel of seismic energy released by earthquakes. The travel times of these packets of energy reveal the structure of the Earth.*

Total 6370 km

## CONVECTION CURRENTS

The mantle convective circulation that dominates the outermost region of Earth (see Chapter 1) has influenced in a singular way Earth's evolution ever since it was born, and because of it the layer on which we stand is often in violent upheaval. The mantle of Earth is in continuous upheaval, since it is being constantly turned over by convection currents, causing the continental plates to shift about on the Earth's surface at the rate of a few to 10 centimetres a year. We cannot feel the steady, minuscule shift—called continental drift—under our feet, but the sophistication of modern astronomical measuring systems is so great that by watching the stars in relation to our own movement, we can actually observe the motion of the planet's surface plates.

The mantle's convection circulation determines the rate at which the Earth cools, by controlling two factors: first the escape of primordial heat from the liquid core; and second the rate at which the heat generated by continuous decay of the long lived radioactive elements—thorium, potassium, and uranium—rises to the surface.

Figure 19 is a model of the main characteristics of the convective circulation as most earth-scientists envision it today. Hot material rises from the core mantle boundary to the Earth's surface, where its melting creates mid-oceanic ridges. Simultaneously cold material descends from the Earth's surface, and where it plunges down it forms deep ocean trenches. The rising hot material and sinking cold material constitute a closed thermal circulation that is manifest at the surface in the pattern of relative motions that drives continental drift.

**Model of earth mantle convection currents**

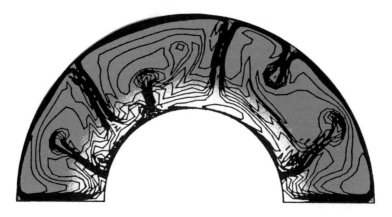

FIGURE 19. *A modern computer simulation of the mantle convection process responsible for continental drift and sea-floor spreading. Hot material rises from the core-mantle boundary to the Earth's surface, where it creates mid-oceanic ridges. Similarly cold material descends from the surface to the core-mantle boundary, inducing deep sea trenches at the surface.*

## CONTINENTAL DRIFT

The radical theory that the relative positions of the continents are not fixed was first suggested in 1915 by the German scientist Alfred Wegener; but it was widely rejected until the 1960s. Scientists now accept that Earth's present continents were once part of a single super-continent they call Pangea. This mass began to break up about 200 million years ago, and the fragments slowly drifted to their present global positions. The discovery of the motion of the plates led to a revolution in the earth sciences that has been continuing ever since. Figure 20 is a recent model of this plate motion pattern, based on seismological and paleomagnetic data. It shows the estimated direction and velocity of the plates.

The earth's tectonic plates move slowly—at present approximately 5 centimetres per year on average. Even so, it is clear that over tens of millions of years or more there will be significant redistributions of the continents. It is because of these slight but continuous rearrangements

## Movement of continental plates

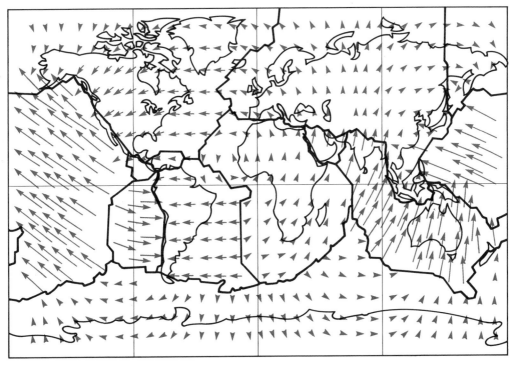

velocity scale: ——▶ 10 cm/year

FIGURE 20. *The speed and direction of travel of the continental plates today, according to a current model of the process of plate tectonics and continental drift. The length of the shaft of the arrows indicates the speed of travel.*

that the mantle convection process exerts such profound control on the physical climate system over long periods.

The first 4.5 billion years of the planet's life very clearly reveals the nature and influence of the mantle convection process that has governed the chemical and dynamic evolution of the planet since it first formed (see Chapter 1). By controlling the drift of the continents over the planet's surface, and by regulating the opening and closing of ocean basins, this process has always determined the nature of the 'field' on which the physical and biological components of the Earth's ecosystem play out their complementary roles.

More than two million years ago the continental masses had been arranged into more-or-less their present positions. At about this same time the ascent of the human species—so beautifully recorded in the fossils of the Olduvai Gorge in East Africa—had clearly begun, and another dominant natural process, glaciation, was underway. By that time Greenland was similarly ice-covered. In the Southern hemisphere, Antarctica was covered by an ice sheet about four kilometres thick, and had been for approximately 40 million years.

In thinking of the Earth as a system, it is useful to see it as composed of many spheres of action, or domains—among them the air, the oceans, and ice. These spheres all operate with different characteristic time-scales: changes in the atmosphere occur most quickly, taking a few years; in the oceans the characteristic time for mixing is a hundred years or so; and in ice, thousands of years. The different spheres of relativity, which together constitute the 'Earth System', fit together in a delicate and interlocking way. If we can grasp the delicacy of this balance, and the nature of the connections between the various planetary changes that involve these spheres, and if we can discern the natural variability of these changes, we may begin to come to grips with global change, or at least make a start in understanding it. If, furthermore, we can filter out the effects of the natural dynamics of the Earth, we may be able to recognize a residual, human-initiated effect. This portion we perhaps have some hope of controlling. This is why we are led to develop more and more sophisticated models of the influences on our planet, trying to see how they interrelate, so that we may better identify the influence of the human component.

## GLACIATION

The story of one of these spheres—ice—of how and why it appears, is remarkable. We have been theorizing about it for over a century, and the quality of our observations improved enormously when we recovered the first sedimentary cores from the deep sea. These observations have revolutionized our thought, and because of the new data we are beginning to discern clearly the way the sphere of ice is controlled by the Sun. The unravelling of this story demonstrates just how precariously the environmental spheres are balanced, one with the other.

To remind us of the most recent history of ice on this planet over the last few tens of thousands of years—just an instant in the age of the

## Ice Sheets

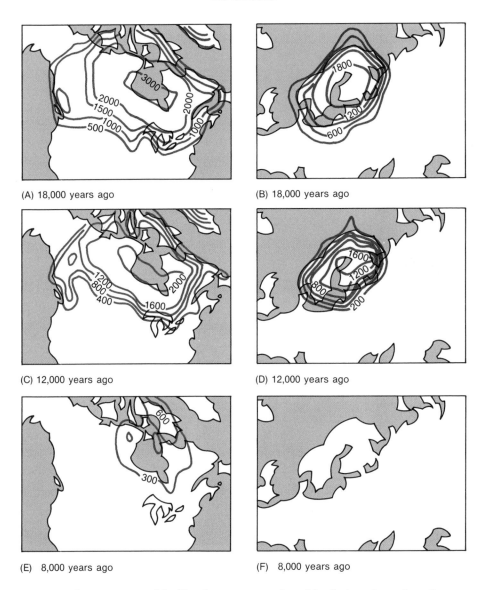

(A) 18,000 years ago       (B) 18,000 years ago

(C) 12,000 years ago       (D) 12,000 years ago

(E)  8,000 years ago       (F)  8,000 years ago

FIGURE 21. *A computer model of ice sheets over northern North America and northern Europe at 18,000, 12,000, and 8,000 years before the present, showing ice thickness in metres.*

Earth—if you had been flying over North America 20,000 years ago and looked down over the North Pole, you would have seen almost all of Canada covered by a huge ice sheet, as in Figure 21. Even 14,000 years ago, this picture wouldn't have been much different. The ice sheet was thick—about three to four kilometres deep, centred geographically over Hudson Bay, and so large that in order to grow it, the sea level had to fall 60 to 65 metres. This phenomenon was therefore no minor event! The ice sheet itself was both an effect and, as it grew, a major cause of climatic change.

The global extent of ice coverage 20,000 years ago involved much more than the huge ice sheet that sat on Canada. Greenland was also more heavily glaciated, as was Antarctica, as was a vast region of northwestern Europe, over the Barents and Kara Seas, the Gulf of Bothnia, Scandinavia—a region so large that it required an additional fall in sea level of about 65 metres to generate the ice coverage.

These ice sheets began to melt about 18,000 years ago, and by 6,000 years ago had almost entirely disappeared, causing the sea level to rise by the same amount that it fell as the ice sheet complexes grew. By that time the Earth's surface had acquired nearly the same distribution of continental ice as at present. From 18,000 years ago until today the sea level has risen close to 130 metres average over the entire global ocean.

## ISOSTATIC REBOUND

Great as that increase in sea level was, there was another modifying process going on at the same time. As the ice on the land melted and the melt water ran off into the sea, the land was released of an enormous load. In response the unloaded regions rebounded and rose. In all the regions that were once ice-covered—Canada, northwestern Europe, western Antarctica, Greenland—much of the land today is well above sea level, forced up by the release of the ice. We call this process isostatic rebound, or isostatic adjustment, and the physics of it is very simple. Over the tens of thousands of years that the ice sheets pressed down on the earth's surface, the 'solid' earth beneath them was eventually depressed as the underlying mantle flowed from beneath it like a viscous fluid. The material under the ice was squeezed out by the weight of the load and flowed horizontally from under it. When the ice melted, the material slowly oozed in the reverse direction and land once covered by ice rose out of the sea. Even today this process continues, thousands of years after deglaciation was complete.

This process has left an indelible record on the global landscape, particularly that of Canada, which had been covered by the greatest

*Ancient beaches in the Rich-mond Gulf of Hudson Bay. Radiocarbon dating of materi-als from the different levels shows the local history of relative sea level change.*

*Raised beaches on a small island off the Melville Peninsula, NWT. Each of the concentric shoreline rings has been produced as the land has slowly emerged from the sea in the process of isostatic rebound. The oldest shorelines are in the centre and highest part of the island, as they were the ones that first emerged. (Photo Alan Morgan.)*

load of ice. In Richmond Gulf on Hudson Bay, for example, the hillside is cut by a series of beaches that from the air look like giant staircases. They were created as the ice began to melt roughly 18,000 years ago; the land began rising higher and higher out of the sea through the process of isostatic adjustment. Each stair is an ancient sea level, a place where the sea used to intersect with the land. Sea level histories like this have been collected from many places on the Earth's surface.

Today we can walk these beaches, collecting the shells of the dead animals that were either stranded out of the sea (if local sea level was falling) or submerged into the sea (if it was rising). The shells of these familiar clams and related species continue to reflect the chemical characteristics of their environment while they were alive. Carbon 14 radioactive dating reveals their ages, so that we may reconstruct a local history of relative sea level change.

These data play an extremely important role, not only in understanding past climatic change, but also in constructing geophysical models of the mantle convection process. It is the best information we have to measure the viscosity of the planetary mantle that limits the strength of the convection process and therefore the rate of continental drift. Isostatic rebound dominates the 'natural' variability of sea level in regions that were ice-covered, and must be taken into account everywhere when we try to measure possible sea level changes due to human activities.

At sites that were once ice-covered, like the Richmond Gulf, sea level has continued to fall at a progressively decreasing rate since deglaciation. Models show that maximum rates of change at present are nearly one centimetre a year. In ring-shaped areas around the centres of postglacial rebound, however, the sea level is steadily rising today at a maximum rate of 2 millimetres a year. At sites like Boston and Halifax, located near the ancient margin of the Canadian ice mass, the record of sea level change shows both rise and fall. Further still from the main centres of glaciation, throughout the major ocean basins, the sea level, with local variations, in general falls at a rate of nearly .5 millimetres a year, due solely to the process of glacial isostatic adjustment.

## GLOBAL SEA-LEVEL RISE AND THE GREENHOUSE EFFECT

It's not surprising that the expected increase of mean-surface temperature due to the release of carbon dioxide and other greenhouse gases is likely to cause rising sea levels (see Chapter 2)—for two reasons. First, increasing atmospheric temperatures will inevitably be passed on to the oceans, which will expand, raising the sea level. We are not

## Tide-gauge records — Locations

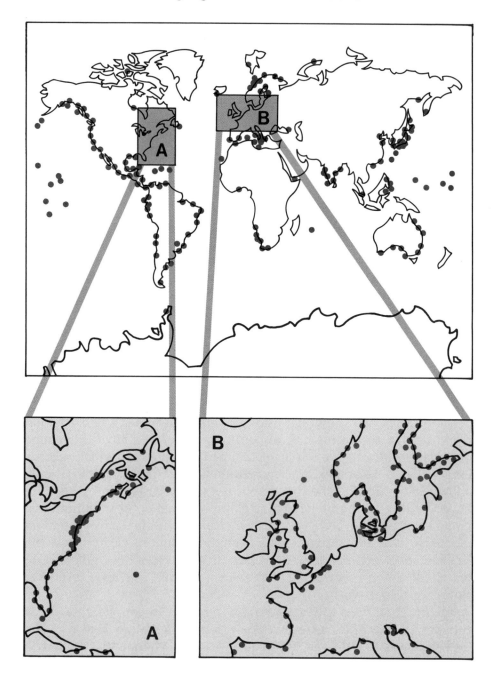

FIGURE 22. *Locations around the world from which relative sea-level histories are available.*

at all certain how rapidly this will happen (see Chapter 6). However, probably more important is a second influence, namely increased ocean volume caused by the melting of continental ice. Since most land ice is located in high latitudes, and since the impact of greenhouse warming will likely be most pronounced near the poles, this cause of rising sea levels will probably dominate.

**Sea Level Trends**

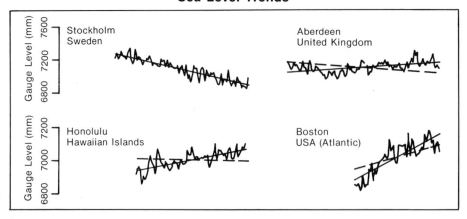

FIGURE 23. *Some long tide-gauge records (solid curves). Where the rate of sea level change would be different if glacial isostatic adjustment alone were the cause, a broken line shows the theoretical rate.*

Do observations show that a component of sea level rise exists that could be due to the melting of continental ice? We have tide-gauge observations of relative sea-level change from sites for which records longer than a decade exist at an extremely large number of locations, shown in Figure 22. (The coverage is not uniform, as most gauges are located either in the Baltic Sea or along the east coast of continental United States, and both regions are strongly influenced by ongoing glacial isostatic adjustment.) Data from these stations clearly show the isostatic rebound experienced by each area. But that does not account for all of the present-day variations of relative sea level. Another process is also at work.

Figure 23 shows the variation of monthly mean heights for a number of examples of particularly long tide-gauge records. The broken lines show the rates of relative sea level change that should be observed, if glacial isostatic adjustment alone were active. The global pattern of this isostatic adjustment signal is shown in Figure 24. It does seem that, except at the Baltic Sea sites, other factors are contributing to the observed changes. Careful analysis of 50-year-long records indeed shows that this residual signal is probably caused by the melting of continental ice and could therefore be due to greenhouse warming. If global sea level rise should accelerate, as is expected, it could have a devastating impact on the very large number of people who live in areas that would be innundated. We may in fact already be observing its beginnings.

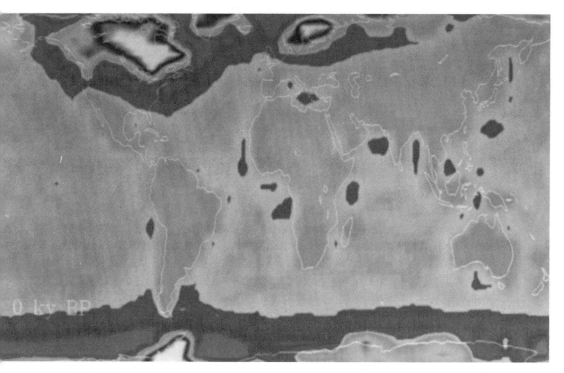

FIGURE 24. *Model of the present rate of global sea level rise or fall, using the long tide-gauge records.*

## CAUSES OF CLIMATE CHANGE

Where did the ice sheets come from in the first place? How often has the planet been glaciated in the past? How likely is this to happen again? It is only in the past two decades that we have begun to answer these questions. The introduction of computerized modelling, along with the analysis of the contents of cores drilled from the major ocean basins, have helped us answer some of them. The sediments in these very long cores, some more than several hundred metres long, provide information about the volume of ice that existed on the continents in the past, and why the ice sheets appeared and disappeared. This story is one of the most exciting in modern Earth science.

Ice sheets are made from snow accumulated and compressed over hundreds or thousands of years, and the water that makes the snow is produced by evaporation off the surface of the sea. However, the water vapour that is precipitated as snow in the construction of an ice sheet is anomalously rich in the lighter isotope of oxygen (the water contains more $H_2^{16}O$ than average), leaving the water in the ocean isotopically heavier than average (containing more $H_2^{18}O$), and this signature is recorded indelibly in the shells of micro-organisms called Foraminifera living in the water. When these animals die, their shells sink to the sea bottom. The proportion of different oxygen isotopes in the accumulated shells in the sedimentary column therefore tells when glacial periods occurred and how much ice existed on the continents through time.

In the record of oxygen isotopic variations in a core from an ocean-drilling project site in the Panama Basin, certain patterns emerge. The record displays a history of about two million years, with a marked transition in character near the mid-point. The earliest half of the record reveals relatively small fluctuations in the proportions of heavy and light oxygen isotopes. But in the last million years the record shows large changes, indicating times of large ice volume on land. The time between minimum and maximum volume is about 100,000 years during this most recent period.

A detailed analysis of this time-series record reveals a number of very well-defined characteristic periods, representing dominant cycles of increase and decrease in heavy-oxygen isotopes, and therefore heavy glaciation. In the last million years a major cycle of cold and warm periods occurred every 100,000 years, the very time-period over which we think the Laurentide Ice Sheet of Canada appeared and disappeared. From this record we can extrapolate that Canada was heavily glaciated roughly ten times over the past million years. There are four subsidiary cycles at about 41,000-, 23,000-, 21,000-, and 19,000-year periods. For the earlier million years of record, all of the four subsidiary peaks are present, but the 100,000-year component does not show.

These characteristic time-periods, and this history of ice-building, were postulated forty years ago by a Serbian physicist called Milutin Milankovitch, who proposed, following earlier suggestions of the German climatologist Vladimir Köppen, that ice ages occurred in response to very small changes, over time in the amount of radiation received by the Earth. These changes were caused in turn by variations in the geometry of the planet's orbit around the Sun (Figure 25). The reason: the geometric properties of the Earth's orbit vary in time, and affect the amount of sunlight reaching the planet. While the average over the year of the net radiance was the same, some summers were a little hotter than usual, and some were less hot or even cool. Although his theory was ridiculed for decades, Milankovitch continued to believe that ice sheets were extremely sensitive to summer temperatures. The winters were always cold enough to sustain an ice sheet if it was already there, but if summers got so warm that the ice sheets melted back, the succeeding winters may not have been cold enough to allow the ice to recover its full volume.

Using an extraordinarily complex formula, long before computers, Milankovitch determined all the tiny changes in the effective sunlight intensity of the summer and winter seasons over hundreds of thousands of years of the Earth's most recent history. Finally, he firmly demonstrated that every one of the recently observed glacial periods—100,000, 41,000, 23,000, 21,000 and 19,000 years (with the possible exception of the longest)—is a characteristic period expected in the ice-volume record due to the variation of one of the Earth's orbital parameters.

## Orbit of Earth Around the Sun

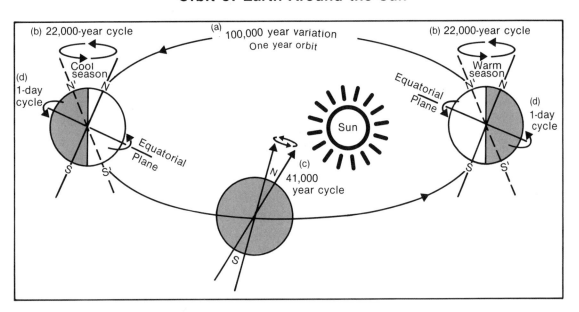

FIGURE 25. *The geometry of the Earth's orbit around the Sun, showing the principal sources of variability.*

What are the varying orbital parameters that control the amount of sunlight reaching us? We all know that the Earth's orbit is roughly elliptical—almost circular. The elliptical shape of the orbit itself, under the same multiple gravitational attractions, also varies, with a dominant characteristic period of 100,000 years.

As the Earth moves in this elliptical orbit around the Sun, its 24-hour rotation gives rise to night and day. At the same time as the

Earth spins around its axis, the axis itself remains tilted at an angle of approximately 23½ degrees to the plane of the Earth's orbit around the Sun. If the Earth and the Sun were alone in the solar system, all the geometric properties of the orbit would be rigidly fixed. Because they share the system with other planets, tiny variations occur. The tilt in the spin-axis in particular is dominated by a characteristic period of 41,000 years. As the non-spherical Earth spins around the Sun— pulled by the gravitational force of not only the Sun but the moon, Jupiter, Mercury, Venus, and Mars—the direction towards which the spin-axis points with respect to the fixed stars also executes a conical motion in space. It slowly wobbles, like a top, completing a cycle in about 22,000 years. In consequence the seasons precess, or come a little earlier over the years, and therefore correspond to successively different positions in the orbital ellipse (Figure 25).

Of course the climate is not directly altered by these small changes in the Earth's orbit around the Sun; climate is sensitive rather to variations in the distribution, and the effective intensity, of the solar radiation received. It's a relatively straightforward matter to reconstruct the history of small fluctuations in the Earth's orbital geometry. In doing this we must include the influence of all the planets in the solar system on each other. With this reconstruction we can compute the history of variations in solar radiation and come to grips with the cause of fluctuations in continental ice-volume.

Some important issues remain to be resolved concerning what has come to be called 'the astronomical theory' of ice ages. Foremost among these is the reason for the 100,000-year cycle seen in the analysis of the last million years of oxygen isotope variability, and not in the earlier million years. Nevertheless we cannot now deny the fact that the climate system does respond dramatically to the extremely small fluctuations of sunlight caused by variations in the Earth's orbit around the sun. The climate system must therefore be a highly non-linear system, perhaps often responding dramatically to small external and internal stimuli.

A physical model that has been used most successfully to describe the link between variations of incoming solar radiation and ice-volume change involves an interaction between several elements of the global Earth system. In this model the characteristic 100,000-year variation that has dominated the last million years of earth history is shown to depend upon:
—the solid earth;
—the ice sheets;
—the oceans and their role in the carbon cycle;
—the interaction between these three elements and astronomical changes in the radiation field.

The complexity of this interaction is a striking example of the way the intricacy and non-lineality of the climate system can produce subtleties and surprises. In response to very tiny changes in the Earth's orbit, altering the heat of summertime seasonal solar energy by only a few percentages, huge ice sheets are created, to grow and decay. What could be more eloquent proof of the extraordinary fragility of the planetary environment?

The recognition that humanity has become an active player in driving the evolution of the ecosphere encourages us in our efforts to understand further the way in which the physical, chemical, and biological components of the Earth system mutually interact. It is only through such improved understanding that we can hope to evaluate the influence of human beings on a global scale, and thus be able to take remedial action where needed.

# CHAPTER 4

# Approaching Today

## JOHN V. MATTHEWS JR

The life-history of the Earth, from the recent past to millions of years ago, is preserved in layers of sediment. They enable us to travel back to the prehistoric eras and reconstruct what a given area was like, how the sea level varied, what animals and plants lived there at different periods, and the incidence of traumatic events such as volcanic eruptions, earthquakes, and catastrophic floods. From this information we can draw conclusions about how human beings long ago used and traversed the land. Moreover, by looking at prehistoric changes in this way, at their nature and rates, we can distinguish between 'natural' and human-caused changes, giving us a base for understanding, even sometimes predicting, and therefore rationally dealing with, what is to come. In other words, we can get a perspective on what the Earth may be like in the future.

So far in this book we have seen that the dynamics of Earth have mostly been very long-term, many millions of years in the making. Now we are ready to focus on changes within the last few million years. Compared with most of the global changes we've considered so far, this is a relatively short term, contained within a fraction of the planet's life.

We can read the story of the changing Earth in the last five million years in a hypothetical sequence of layers exposed in a steep river bank about 30 metres high. It is located in Canada's northern Yukon. Containing the features we need to reconstruct our past, it is a typical northern river bank, which could just as well be in Alaska or Siberia. The river is a torrent in the spring during snow-melt, and this tends to keep the face of the bluff fresh by washing away portions that have become overgrown. Our river bluff is located within the permafrost zone, where the sediments are permanently frozen below a few tens of centimetres. A map of soil temperature and vegetation shows that the regional treeline, about 50 kilometres south, coincides with the northern limit of subarctic soil temperature, with a mean of 13 degrees centigrade in July (Figure 26). Although treeless, the top of the bluff is covered by a rich variety of shrubs and herbs.

We first see the site from the boat in which we ascended the river. On our way upriver we passed many sections that seemed similar, but past experience, and a practised eye, show that this one contains the most complete geologic record. For instance, we can see that the basal sediments near the water level have large pieces of wood—larger than trees now growing just south of the site. From the river we also see that the section contains glacial sediments. The abundance of organics in most of the sediments—revealed by their dark colour and by the presence of actual peat layers—also means that this site holds the greatest promise for finding fossils that will tell us about ancient climates. Finally, close inspection of the cliff with binoculars reveals the 'prize'—a thin line of white volcanic ash—that makes this section more important than all the others we have seen.

*The site of our geologic 'walk-down', containing fossils, sediments, volcanic ash, and other features that can tell us about climatic conditions at the times the layers of the bluff were deposited. Layers of the river bluff, which we approach by boat, hold a complete geologic record of the last five million years.*

**Soil temperature and vegetation today**

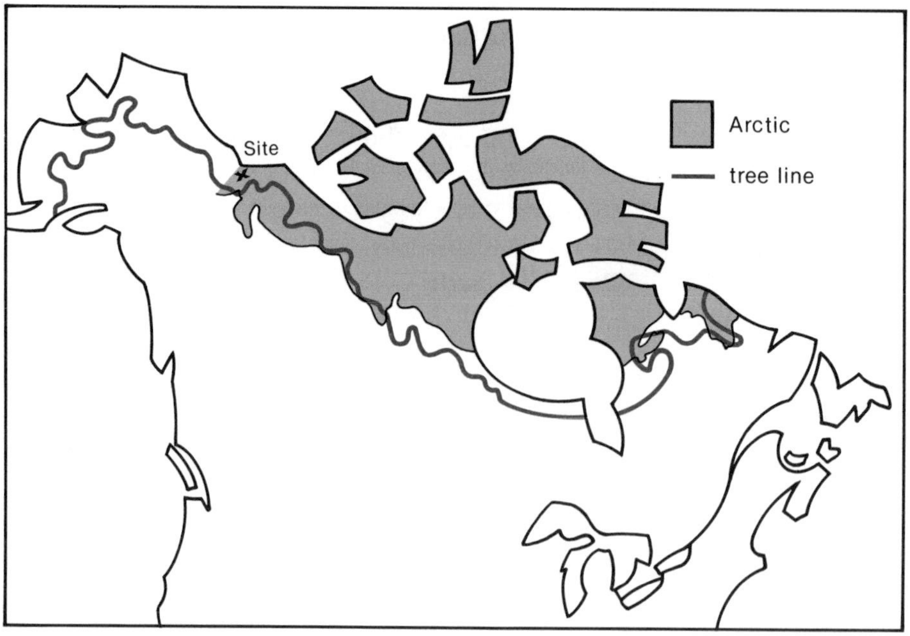

FIGURE 26. *Soil temperature regions of Canada, showing treeline and location of site.* (*From* The National Atlas of Canada.)

Normally when studying such a site, a geologist would work from the bottom up, describing each layer of sediments and collecting sample fossils or sediments to analyse later. We, however, will look at the consecutive stratigraphic layers in reverse order, walking down the bluff and back in time. Because we are trying to make sense of what we see on the spot, we will be considering what scientists call 'proxy data', substitutes for actual observations of a past event. Most of our proxy data are derived from fossils, sediments, and other features on our bluff, but we will also take into account some written records of the past, and geological knowledge of other areas. All together, this information can give us a good idea of conditions, including the climate, at the time the layers of the bluff were deposited.

Before beginning our 'walk-down' in geological time, we stand at the top of the bluff for a moment, surveying the view of the lake-dotted basin. That view has changed over time, showing many different faces. Our task is to identify them and decipher their story of ancient climate regimes.

*This lake-dotted river basin holds the story of ancient climate regimes on the planet Earth. It is within the permafrost zone, where sediments are permanently frozen below a few tens of centimetres.*

## LAYER ONE—THE PRESENT PERIOD

The layer we are standing on is made of peat, two or three metres deep, that accumulated after the ice of the last continental glaciers retreated, about 10,000 years ago. Since it is the youngest part of the riverbank we are to probe, it contains the best-preserved fossils and the most detailed record of recent climate changes. There is a forest-bed buried near the base of this layer, and individual trees can give us a tree-ring record of climate change. Their woody cells preserve a detailed life-history of the individual tree: its age, rate of growth, response to environmental change, disease, and damage—even the isotopic composition of the rainwater at the time the tree was alive.

Carbon 14 dating of the wood shows that these trees grew about 10,000 years ago. The presence of the forest bed alone tells us that the climate was warmer then, able to accommodate the growth of trees further north. When we scrape away at the surface of the peat, we discover that the sediments are frozen. If we were to bore a hole right through this layer, and measure the temperatures at different levels, those measurements would give us a record of the most recent climate change. Temperature change at the surface creates a response that travels down into the earth, so that the surface-temperature history of the past 100,000 years is contained in the upper 1,500 metres of the Earth's crustal layers.

The top few centimetres of peat contain not only organic material, but also the remains of a former Hudson's Bay Company post, a busy trading centre active on this spot 200 years ago. Part of the job of post superintendents was to make climatic observations, and many of them also recorded proxy climate information such as time of flowering of plants, migration of waterfowl, and the pelt quality of fur-bearing animals. (Such proxy climate information was also recorded in areas with long historic traditions, like China and Europe.) In Canada, even with its relatively short written history, the Hudson's Bay Company records give us valuable information about the Little Ice Age, an interval of relatively cold climate that occurred between AD 1400 and 1850. One of the extreme 'events' within this interval was the 'year without a summer' that followed the 1815 eruption of the Tambora volcano in Indonesia. Hudson's Bay Company records show that in both 1816 and 1817 parts of northern Canada were much colder than the average temperatures today. Food supplies dwindled and grain prices soared as harvests further south failed. This cold was attributed until recently solely to Tambora, but detailed studies of Hudson's Bay archives show that temperatures had been falling since 1809, probably related in some fashion to change in sunspot activity. Tambora's eruption only sharpened the decline.

Layer One tells us only about the last 10,000 years—a long time in human terms, but a short time geologically, and too short to document all aspects of former climate. If we are to achieve a true perspective on how climate has changed in the past, we must walk further down the bluff and examine successively older sediments.

## LAYER TWO—GLACIAL DEPOSITS

We will have to scramble to find a footing to examine the second layer, which is a nearly vertical cliff about five metres high, composed of a very stony, concrete-like deposit known as till. This type of deposit is peculiar geologically because it contains an unsorted mixture of very fine sediments such as silt, and larger stones, some of boulder size, left when the glacier that covered the area between 10,000 and 50,000 years ago melted.

This till is especially significant because it contains rocks called erratics, which came originally from thousands of kilometres away. They show that the glacier which transported and deposited them was of continental proportions, probably part of the last major continental glacier to cover North America—the Laurentide Ice Sheet. We now know a great deal about the expansion and contraction of the Laurentide Ice Sheet from other sites in North America. The area we are studying is located on what was its margin.

Even though we are living today in a warm period, a considerable volume of glacial ice still remains on the land. Large ice sheets still exist in Antarctica and Greenland, and there are also smaller ice caps in the Canadian Arctic archipelago. Since fallen snow and ice traps air, atmospheric dust, pollen, aerosols, acid droplets, and a whole range of exotic chemical products of terrestrial and, more recently, human origin, these ice bodies contain a unique record of our changing world. They are accessible mainly by detailed study of ice cores obtained by drilling through the ice with auger-like tools.

## LAYER THREE—INTERGLACIAL DEPOSITS

As we come to the next lower geological layer, about ten metres thick, we see that it is different from the overlying glacial deposits because of its high organic content. Most of this is peat, formed by continuous burial of surface vegetation. It contains insect fossils as well as plant fossils such as pollen, leaves, and seeds, which do not occur as far north today (Figure 27). We can therefore again assume that this layer was laid down during a warmer climate than at present. What we have found, then, is evidence of a relatively brief warm interglacial period, like the period in which we live today. We know that such interglacial periods happened approximately every hundred thousand years over the last million years. By comparison with the marine record of shifts in isotopic content of sea water, we know that the interglaciation recorded in Layer Three occurred approximately 125,000 years ago (see Chapter 3).

During all interglacials, less water is trapped on the continents in the form of glacier ice. Accordingly, the sea level is inevitably higher than at other times of colder climate. At some sites in north coastal Alaska or Siberia, a layer of the same age would contain a thin clayey stratum with marine shells. Marine sediments would show that the world sea level was approximately six metres higher than it is now.

## LAYER FOUR—COLD CLIMATE, BUT NO GLACIATION

We now encounter a geological unit that is also about ten metres thick. Because it lies below Layer Three, we know it must be older than 125,000 years. Now permanently frozen, it is composed mostly of silt with peat, sometimes formed into vertical wedge-shaped structures one to two metres across, which indicate the former existence of

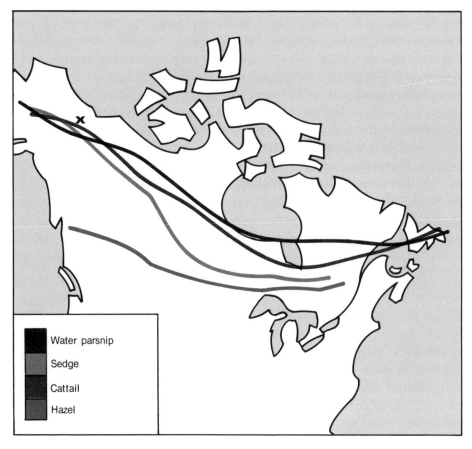

FIGURE 27. *Present northern limit of plant species found as fossils at the site described.*

permafrost that then melted, probably at the beginning of the accumulation of the interglacial deposits in Layer Three.

Though not as cliff-like and spectacular-looking as Layers Two and Three, this layer contains fossils that clearly record the past. Often sediments like this contain mammal bones and carcasses excellently preserved because they have been entombed in permafrost. There are abundant bones of mammoths, bison, and horses, but it is the rarer fossils that provide interesting insight into past environmental conditions. For example, the Saiga antelope skulls found in this region tell us that these tiny skittish antelopes once roamed the areas of northern North America and eastern Siberia that were not glaciated—the glacial refugia. These lowlands cannot have been as boggy and wet then as they are today, and in winter they must have been covered by less than 20 centimetres of snow, since today saiga live only in south-

*There are many remains of mammoths in the bluff. This baby mammoth, 'Dima', was found in 1977 by Siberian gold miners who were thawing the permafrost on a tributary of the River Kolyma. The completely frozen carcass measured 115 cm long and 104 cm high and was probably six to seven months old at the time of its death. It is similar to a baby elephant, but has long hair and very small ears, characteristics portrayed by early cave artists in Western Europe.*

central Asia, in herds that rely on flight over firm terrain to escape predators. The Saiga antelope was not the only animal of its type to live in this area while the silts were forming. It was part of a unique assemblage of animals and plants that existed in other glacial refugia of the world, and have no modern counterparts in those areas.

Many of the large mammals that lived with the saiga, such as the woolly mammoth, the giant bison and sabre-toothed lion, are extinct today. Some researchers believe that they may have succumbed to over-hunting by the first human immigrants to North America. Indeed, these periods of severely cold climate, though they seem hostile to us, were probably quite friendly for our ancestors. The large game provided a huge protein resource. The dry climate and terrain made travel easy, as did the fact that because of lower sea levels, continents and islands now separated by marine barriers were connected. (One of the most spectacular examples of a now-vanished barrier is the Bering Land Bridge, which joined northwestern North America and Siberia, and on which the first people travelled to North America 30,000 to 15,000 years ago.)

As the map of the Laurentide Ice Sheet (Figure 31) shows, some parts of Canada remained ice free during the last major glaciation. The area of these glacial refugia varied from one glaciation to another. Apparently the area we are examining was near one of these refugia, for the only glacial deposits in the entire section are in Layer Two.

This does not mean that the climate was not cold when Layer Four was deposited. Glaciers were advancing elsewhere in North America. In fact the fossils and sedimentary features we find here show just how cold it was. The wedge-shaped structures are 'fossils' of former ice-wedge polygons that are so common in the North and are responsible for the ground-pattern we see when flying low over the Arctic.

One of the deposits in Layer Four is a pearly white seam of volcanic ash. Called tephra, it consists entirely of tiny fragments of glass, formed at the time of an explosive volcanic eruption. Tephra can be dated and thus used to pinpoint the time of eruption. Seldom is it possible for geologists to document such 'instants in time'. Furthermore, because the ash from individual eruptions is chemically distinct, it is often possible to recognize the same type of tephra and hence the same 'instant' at a number of sites. When sediments and fossils found with the tephra are studied, it becomes possible to determine climate and environment conditions at the same time over a wide region. In this case, the tephra tells us that the eruption was much larger than the Tambora volcano in Indonesia in 1815, but probably not as big as the great Toba eruption on Sumatra 75,000 years ago, which deposited ash in India, 3,100 kilometres away.

*A pearly white seam of volcanic ash, called tephra, consists entirely of tiny fragments of glass, formed at the time of an explosive volcanic eruption.*

## LAYER FIVE—PRE-GLACIAL DEPOSITS

We have now reached a level one-to-two metres immediately above the river level. The sedimentary content is mostly sand, but it also contains organic materials like those usually carried and deposited by streams. We do not know how far it continues below the surface of the river, but it could be only the very top of a huge thickness of buried sediments extending far back into geologic time. As it is, we can see that a great deal of time, possibly as much as several million years, passed between the deposition of this layer and the one above it.

Rare beds within this layer preserve fossil tree trunks still erect as when growing. This forest-bed is very different from the one we saw near the top of the bluff, in Layer One. That 10,000-year-old forest consisted of trees and other plants that now live only a few tens of kilometres south. In contrast, we are now looking at fossils of plants such as five-needle pines, honeysuckle, and mulberry, all of which now only grow thousands of kilometers south (Figure 28). Some of the seeds even represent plants that are extinct in North America except where artificially cultivated. Their seeds were deposited here when the climate was not only much warmer, but possibly different in other ways from that found anywhere in the world today.

The plant fossils here are approximately three million years old, predating the earliest recognized glaciation of North America. Some have called this the last warm period before the ice ages, a climatic 'Golden Age'. The only large ice masses in the world were in Antarctica, and they were much smaller than today. Permafrost probably existed only in the highest mountain ranges, in Antarctica and the northernmost fringes of Asia and North America. Greenland was truly green; there was no ice-cap and forests grew almost to its northern shore. All northern coastal regions bordered an Arctic Ocean that was not perennially frozen as it is today, and to a degree it was also a more isolated ocean because of the land-bridge between Siberia and Alaska and a submarine ridge—also once a land-bridge—between Greenland and Europe. Such an ice-free, relatively isolated Arctic

**Distribution of 5-needle pines in Canada**

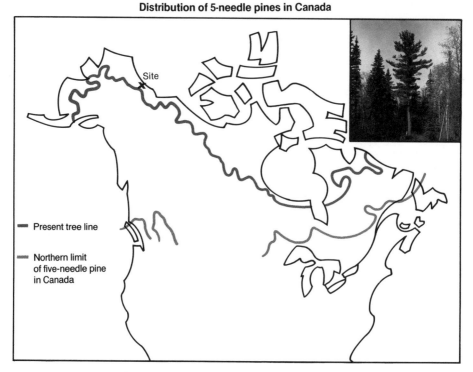

FIGURE 28. *Distribution of five-needle pines in Canada, compared to present tree line.*

Ocean must have affected the flow of heat in all the oceans of the world, and have had major indirect effects on the terrestrial climate of the time.

The pre-glacial climate indicated by this lowest layer is our baseline for comparison when we discuss the climatic changes that occurred during the ice ages of the last 2.4 million years. The archetypal boreal forests that were growing while these sediments were accumulating developed over a long period of time and were not, like present forests, subjected to the disruptions wrought by cold-warm, glacial-interglacial fluctuations.

Since some projections of future climate call for climatic conditions as warm as when the forests of Layer Five were growing, we must probe back at least that far to understand how drastic the future climatic changes caused by mankind might be.

What conclusions can we draw from the information we have gathered in our walk through the past? We have seen that the climate fluctuated—from hot to cold, at least several times in the few million years we traversed. From what we know about the cause of glaciations, we suspect that our current interglacial period, the time in which agricultural and industrial man emerged, will be a relatively short warm interval. If human intervention does not change the pattern, it will probably last not much longer than a few thousand or perhaps even a few hundred years. As we saw in Layers Three and Five, there were other similar warm intervals in the past 1.8 to 2.5 million years (see Chapter 3).

The onset of our present warm period, like other interglacial periods, seems to coincide with particular conditions of the earth's orbit and tilt, but as Figure 29 shows, other changes were occurring at the same time, and it is difficult to know which are causes and which effects. The warming during the present period, as in earlier periods, has been associated with an increase in atmospheric carbon dioxide, and a rise in sea surface temperature.

As we saw in Chapter 3, periodic changes of the earth's orbit are major causes of the climatic changes of the ice ages. These variations can be predicted, and this allows us a large degree of certainty in forecasting the major trends of future climate. But what of unpredictable events, such as volcanic eruptions? We have already seen that the Tambora eruption in 1815 probably helped lower the temperature over a large area in the following year. What if several closely spaced Tambora-like eruptions occurred at a time when the long-term trend of climate was one of cooling? Could they force a rapid climate jump into a glacial mode? The answer is that we don't know—yet. We must look beyond the historic record of volcanic eruptions and into the

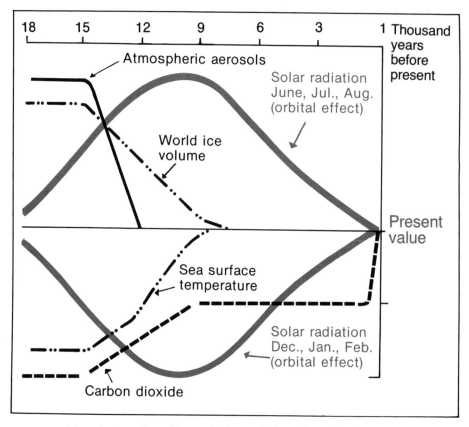

FIGURE 29. *Many factors affect climate. This graph shows how they have changed over the past 18,000 years. As sea temperature rose and achieved its present temperature about 9,000 years ago, world ice volume declined. Atmospheric aerosols declined to the present level about 12,000 years ago, while $CO_2$ rose to a plateau about 9,000 years ago and then rose rapidly in the last few hundred years. The blue line shows the calculated change of the major influence: summer and winter solar radiation received on earth. (Source: J. Kutzbach and H.E. Wright, 'Simulation models of Climate at 18,000 Year BP', Quaternary Science Review, 4, 1985.)*

realm of proxy climate indicators (Figure 30) to discern the way the largest of these unpredictable events influenced climate.

Keeping in mind these variables, some scientists suggest that the climate of the next 200 years, influenced by rising levels of carbon dioxide, methane, and other such gases, will be much warmer than today, similar to that long-ago climate we reconstructed in exploring the bottom layer of sediments in our river bluff. But those sediments accumulated over millions of years, and the changes they represent were very slow. We are contemplating now a rapid change, within the lifetimes of our children. A change of this magnitude, over so short a time, would not reproduce the environments that existed several million years ago, but the fossils from those sediments and others about the same age throughout the world help us get a sense of the possibilities. As in the periods we have explored, we could have climate variations great enough to alter radically vegetation and animal life, to innundate some coastal areas and to melt the ice from others. And, of course, in the future it will be human beings who not only influence, but also suffer the effects of, these global changes.

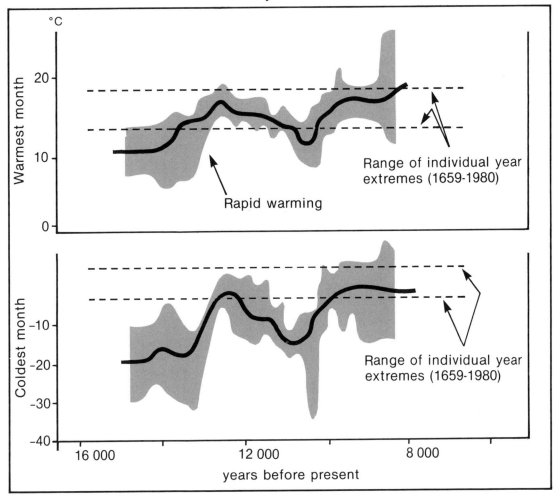

## Local temperature
### Indicated by fossil beetles

FIGURE 30. *Fossil insects reveal mean temperatures in centigrade for warmest and coldest months, from 15,000 to nearly 8000 years before the present. They show rapid warming beginning about 13,500 years ago, a lapse into cold temperatures 10,000 to 11,000 years ago, and the warming trend that has continued today. (Modified from T.C. Atkinson, K.R. Briffa, and G.R. Coope, 'Seasonal Temperatures in Britain During the Past 22,000 Years . . .',* Nature, *Vol. 325, 1987.)*

# The Last Glaciation:
# The Laurentide Ice Sheet

At its maximum, about 18,000 years ago, the Laurentide Ice Sheet influenced the vegetation, drainage, and climate of the whole of North America. It was the last major glacier to advance across the continent, eventually covering almost all of the northern half, from the Atlantic Ocean to the Rocky Mountains, and from south of what is now the Canada-U.S. border to the high Arctic islands. Mountains and plains alike were covered. (Figure 31.)

The Laurentide Ice Sheet began when snow that had fallen on the large plateaus of first Baffin Island, then Quebec-Labrador, and later Keewatin failed to melt during several successive summers. As more and more snow accumulated, it changed to ice and the ice began to spread in the typical fashion of a glacier.

During the early growth of the glaciers, between 120,000 and 80,000 years ago, the ice advanced and melted back several times, but large areas of Canada remained unglaciated. Then a major build-up of ice began peaking about 20,000 years later. The ice margin retreated and fluctuated for another 40,000 years until the start of the final expansion, about 20,000 years ago.

When the Laurentide ice-mass was at its largest, the Greenland, Eurasian, and Antarctic ice sheets were also close to their maximum, and between them the volume of water stored on land as ice was enough to lower ocean levels by about 100 metres.

Deglaciation began shortly after 18,000 years ago, but was very slow for another 5,000 years. A sudden jump in the rate of ice-melt occurred about 8,000 years ago, when the sea penetrated into Hudson Bay and destroyed the core of the ice sheet by breaking it up into millions of icebergs.

While it is tempting to view the advance and retreat of major ice sheets as a proxy of climate, an ice sheet is actually a very poor indicator of climatic change, because the glacial response lags behind climate change by thousands of years. Also, once it has formed, the ice sheet creates its own climate to a large extent, by reflecting solar energy back to space. Consequently, it takes much less of a change in climate to 'grow' an ice sheet than to melt one. This is an important consideration, because even though in our scale of time the process of glaciation is extremely

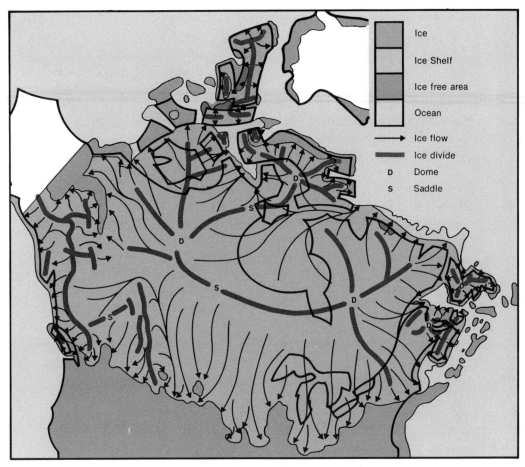

FIGURE 31. *Glaciation 18,000 years ago: the extent of the Laurentide Ice Sheet and flow lines.*

slow, the change in climate that might initiate the start of a slow build-up of ice could be both subtle and rapid.

Sudden warming induced by human activity may create drastic irreversible effects of another kind. For example, as a result of climate warming, the sea level would rise, possibly by several metres. This in turn might be enough of a change to trigger decay of the large West Antarctic Ice Sheet, causing an even higher global sea level, and submerging many of the densely populated coastal areas of the world. This process illustrates one of the most important observations about major glaciers: it takes many thousands of years for a glacier to grow, but it can melt very quickly.

ARTHUR S. DYKE

# Ice Cores

Even in the most pristine environment, fallen snow traps atmospheric dust, pollen, aerosols (fine haze, like fog or smoke), acid droplets, and a whole range of exotic chemical products of terrestrial and, more recently, human origin. It also contains old carbon dioxide in bubbles within the ice, and the water in the ice displays differences in content of two oxygen isotopes.

Above a line on a glacier or ice cap, known as the 'firn line', snow never melts completely. Instead it slowly compacts to form ice—which includes many of the contaminants in the original snow, some of which are valuable proxies of climate, or of catastrophic phenomena like large volcanic explosions; others document human pollution, such as arctic haze. An accumulation of glacier ice may then contain a time-series of historic climate data stretching thousands, even hundreds of thousands of years, in the past.

Cores from present ice caps in Greenland and the Canadian Arctic span about the last 130 thousand years—an entire glacial cycle (Figure 32). A recent core from Antarctica encompasses even more time, including the last interglacial period and the end of the glaciation that preceded it. This 2,083 metre-long ice-core was recovered over a period of several years at the Soviet station of Vostok, in East Antarctica, which has an altitude of 3,488 metres. A Franco-Soviet team has recently completed an analysis of the carbon-dioxide concentrations found in the Vostok ice core. Its recovery and analysis are truly significant because the information it contains extends back 160,000 years, through the last interglacial period to the earlier ice age (Figure 33).

Analysis of the air trapped within the Vostok ice core showed that temperatures during the last interglacial period, about 125,000 years ago, were two degrees warmer than at any time since the ice sheets began to melt 18,000 years ago. In fact, at its peak, the world climate of the last interglacial period may have been similar to what is predicted for our 'greenhouse warmed' future.

The carbon dioxide bubbles trapped in the core show a wide range. The high level is comparable with the pre-industrial level of roughly 200 years ago, and the low level is among the lowest values known.

Analysis of the layers in the ice cores from Arctic Canada show

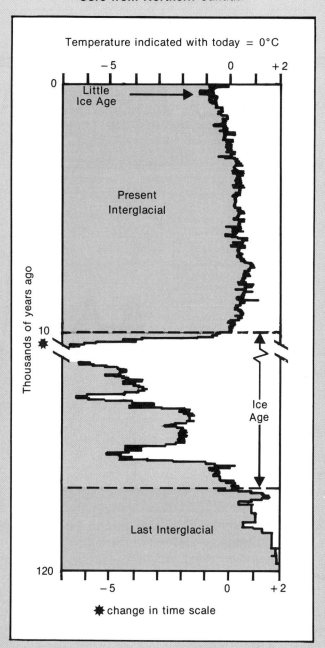

FIGURE 32. *Oxygen bubbles trapped in an ice core from the Agassiz ice cap in Canada's Northwest Territories show temperature fluctuations over 120,000 years.*

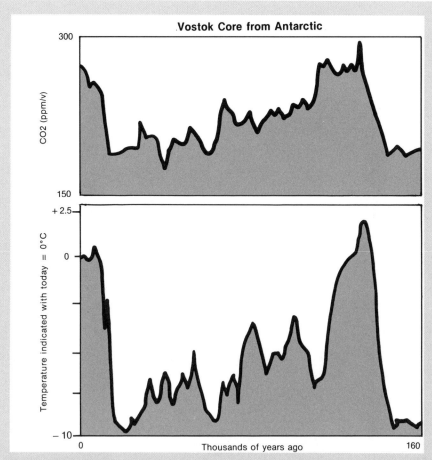

FIGURE 33. *Elements trapped in an ice core from Vostok, Antarctica. Above, carbon dioxide, below, deuterium, a proxy for temperature fluctuations over the past 160,000 years.*

that the summer temperatures were warmer by about 2°C at the beginning of the present interglacial period, 10,000 years ago. However, by 2,000 years ago they had cooled to about 1°C colder than now. So when we think of the present as warm, it is by comparison with the last 2,000 years, but not with the last 10,000 years.

This profile of changing carbon-dioxide concentrations represents a major advance in our understanding of past climate change. It suggests that changing concentrations of carbon dioxide in the past acted to amplify rather than cause climate changes. If the rise of atmospheric carbon dioxide now occurring has a similar influence, it will modify, rather than cancel, ongoing climatic oscillations. R. KOERNER

# PART III

# *A Closer Look*

In June 1988 a world conference was held in Toronto, 'The Changing Atmosphere: Implications for Global Security'. That such an international meeting was held shows us that our political leaders and scientists are worried about our common future. No nation controls its atmosphere. We are all linked together by the atmosphere-hydrosphere system—we do not and cannot live in isolation.

The Toronto meeting was closed with the statement: 'Humanity is conducting an unintended, uncontrolled, globally pervasive experiment whose ultimate consequences could be second only to a global nuclear war . . . It is imperative to act now!'

All scientists who study and predict climate and examine what we are doing today to our atmosphere agree that the first major impacts will appear at high North-South latitudes. Thus Canada, with its vast northern land areas, is at centre-stage in the unravelling of the record of major global change. In this section we will examine parts of the recent record of change. We will look first at what the polar ends of the Earth, with their particular sensitivity to climate change, can teach us about the past and possibly our future. The oceans, themselves tempering our atmosphere, and the lakes, rivers, and wetlands, have their own story of change to tell.

Finally, forests and grasslands clearly show us the results of human actions and at the same time have an input of their own into the equation of global change.

WILLIAM S. FYFE

# CHAPTER 5

# The Polar Regions

## FRED ROOTS

The Arctic, the lands under the constellation Arktos or the Great Bear, has from the times of classical Greek mythology aroused in humans a sense of wonder and mystery. Even the most commonplace and dependable acts of nature, such as the rising of the Sun in the morning and its setting in the evening, were strange and different. In the North, in the land of the Great Bear, beyond the Arctic Circle, the Sun went round and round in the summer without setting, and at the time of the winter solstice it did not rise at all. The sea froze, great mountains of ice floated on the ocean, curtains and shafts of strange lights danced in the heavens, and the mariner's lodestone spun end-lessly on its thread without settling to point toward Polaris, the North Star.

Today scientific knowledge provides an explanation for many of the natural mysteries of the arctic region. But this knowledge only rein-forces the wonder and fascination they evoke. The polar regions, both north and south, have a special significance in the workings and the nature of the planet of which all humans are a part. The response of the arctic regions and their inhabitants to changes provide, if we can learn to read the signs correctly, a sensitive indicator or warning of possible changes to our global home.

## THE PLANETARY SETTING

The distinctive features of the arctic regions are direct consequences of the shape of the Earth and the tilt of its axis with respect to its orbit around the Sun, the electromagnetic dynamics of the spinning planet, and the shapes and asymetric distribution, on a grand scale, of the continents and ocean basins (see Chapter 3).

When, in what people in the northern hemisphere call summer, the north end of the Earth's axis of rotation tilts towards the Sun, arctic

*Beyond the Arctic Circle the sun goes round and round in the summer without set-ting. This multiple exposure was taken on Ellesmere Island at 20-minute intervals from 12:30 to 2:50 a.m. in July. (Photo George Hobson, Polar Continental Shelf, EMR Canada.)*

*Great mountains of ice float on the ocean. This iceberg is on the west coast of Green-land. Most icebergs go north along the Greenland coast and south down the Canadian coast, at considerable hazard to shipping. (Photo George Hobson, Polar Continental Shelf, EMR Canada.)*

regions receive radiation continously; when the axis is tilted away from the Sun in winter, there are periods, lenghtening as one gets nearer to the Pole, when the land and the ocean are continuously in the Earth's own shadow. However, the polar axis is always highly inclined to the Sun, which is therefore always quite low in the sky in the polar regions. The result is a net annual loss of energy or heat from arctic regions. This loss is balanced by the transport of heat toward the poles by ocean currents and atmospheric circulation. Even so, average temperatures are several tens of degrees lower than those in temperate latitudes. The result is ice on the ocean, prolonged snow-cover on land, and on sea ice and glaciers; these white surfaces in turn reflect back rather than absorb much of the already small amount of solar energy received. Because these processes and conditions operate and react at different speeds, any disturbance in one of them can have complex and prolonged repercussions on the environment of arctic regions.

The large-scale architecture of the Earth, the distribution and shapes of continents and ocean basins, channels the ocean currents and influences the pattern of atmospheric circulation. Heat is transported from temperate regions into the Arctic mainly in the eastern North Atlantic, and sometimes very cold air from the Arctic may be carried into the central parts of North America or Eurasia, as anyone who lives in Saskatchewan or Siberia can testify.

All these phenomena cause profound differences in the arctic environment from place to place. The ground is almost treeless and perenially frozen near Hudson Bay, at the same latitude that magnificent stands of oak trees thrive in England and tulip fields bloom in Holland. Rich, commercially productive fishing-grounds are open throughout the year in the Barents Sea near the south coast of Spitzbergen, while at the same latitude Prince Gustav Adolf Sea, in the Canadian archipelago, has not been free of ice in historic times, and its perennially dark waters support only a very few specially adapted fish.

If we look at a globe we can see one reason why the environment of high northern latitudes is particularly sensitive to global change. The polar regions, where more heat is lost to space than is received from the Sun, are smaller in total area than the tropical regions, where solar-energy input slightly exceeds terrestrial-energy loss. The area within the Arctic and Antarctic circles (66½ degrees North and South) is only 20 per cent of the area between the Tropics of Capricorn and Cancer (23½ degrees North and South). Thus, in the transfer of energy from the tropics to the poles, small temperature changes in the tropics lead to larger changes in polar regions. The arctic environment is therefore very dependent on the routing and efficiency of north-south heat transfer, and mechanisms for redistribution of energy around the Arctic itself are poor. Modelling of oceanic and atmospheric responses on a global scale (Chapter 7) predicts quantitatively how changes in ocean temperature or circulation, atmospheric circulation or cloudiness in other parts of the world, will cause a reaction, often exaggerated, on the Arctic and its environment.

What about 'the other end of the Earth', the Antarctic? Although it experiences the same general characteristics of solar radiation, heat loss, and electromagnetic anomalies as the Arctic, the result is in many ways quite different because of its different architecture and dynamics (Figure 34). The broad ocean that surrounds the central Antarctic continent is part of the world's main ocean system, whereas in the Arctic a small ocean is surrounded and confined by large land masses. The most vigorous coupled atmospheric-ocean circulation anywhere on the planet dominates the sub-Antarctic in year-round storms—the 'Roaring Forties', the 'Furious Fifties', and the 'Screaming Sixties' of mariners' tales testify to the furies at these southern latitudes. This vigorous west-to-east circulation effectively mixes and distributes the energy being carried from the tropics to the southern polar regions. The Antarctic continent itself has accumulated snow over millions of years until it is covered by a continental glacier, in places more than four thousand metres thick, over which the cooled air descends and flows radially and spirally to the coast and over the adjacent ocean, freezing the sea surface, creating a broad but annually fluctuating belt of sea ice, anchoring a single great circulation pattern, or 'vortex', that gives a simplicity of structure to the Antarctic climate and environment that is not found in the Arctic. This also results in an

**Arctic Ocean and Surrounding Land**    **Antarctica and surrounding ocean**

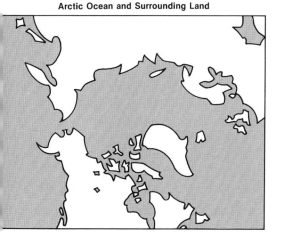 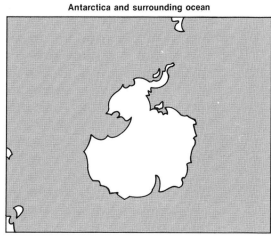

FIGURE 34. *The Arctic and Antarctic. The Arctic region consists of a small ocean surrounded and confined by large land masses of two continents, whereas the Antarctic is a continent itself, surrounded by the world's main ocean system.*

environment on land that is more harsh to life than any other place of comparable size, but a sub-Antarctic oceanic environment that is very productive. The large scale of its environmental processes, the enormous thermal inertia of the Antarctic ice-cap and of the Southern ocean, and the more brutal life conditions make the Antarctic less sensitive to climatic and environmental perturbations than the Arctic. Yet these same characteristics, with the virtual absence of alteration by humans until very recently, make it unique as a demonstration of global environmental history and change. By observing the similarities and differences in the Arctic and Antarctic, we learn much more about planet Earth and its inhabitants than if we were to study just one end of our round planet.

## ARCTIC AND POLAR ECOSYSTEMS

Polar ecosystems must adapt to low-energy conditions. The species must conserve energy and chemical flow, and be able to put on spurts of growth or population expansion to take advantage of favourable conditions. Among their challenges is the need for liquid water, and in a region where most of the time the water is not liquid but solid (snow and ice in various forms), special adaptations are required. Another problem is the adjustment of life-cycles to the time and intensity of solar-energy flow. Because the yearly arctic surface temperature fluctuates above and below the freezing temperature of water, most of the energy received or given out at the surface is taken up in producing changes of phase—to thaw or freeze or make frost crystals—rather than in changing the temperature. At the peak of solar energy, in late June, when plants and animals in other parts of the hemisphere are growing most vigorously, arctic lands and waters are still shrouded in snow and ice. In some places August is well advanced before the ice and snow have disappeared, and plants and animals must carry out the summer portion of their annual cycle under conditions of declining solar energy. Moreover, at low temperatures, and without liquid water in soils and rocks, the rate of chemical weathering is very slow, and minerals tend to break up physically without releasing the soluble chemical nutrients essential to healthy biological growth. At sea the yearly freezing of the surface, which produces nearly fresh ice from salt water, produces a 'rain' of excess salt that sinks to lower depths, scrubbing out many of the nutrients from the upper, comparatively warmer, layers of water. In general the vigour of biological life in arctic regions tends to be controlled not so much directly by low temperatures as by lack of nutrients.

To cope, arctic organisms and ecosystems have developed some unusual practices. Some plants can mobilize chemical reactions in the presence of vapour from snow and ice crystals at temperatures well below freezing, and, aided by sunlight that penetrates the snowpack, start photosynthesis and be already 'green' when the snow melts. Other arctic plants, invertebrates, reptiles, and fish have special 'anti-freezes' that can congeal without developing destructive crystals, so cell tissues can freeze repeatedly without damage. A number of species can interrupt their growth or reproduction cycle and carry on when energy returns or is accumulated, as in flowering plants that may take several short summer seasons to progress from bud to seeds, or muskoxen that may not breed for several years.

In the Antarctic, where living conditions are even more extreme, the adaptations may be still more unusual and fascinating. The persistence and patience of the large emperor penguin, which breeds on the coastal ice, intrigues nearly everyone. Lacking the place or materials to make a nest, it holds its single precious egg on top of its warm webbed feet, protected by a flap of stomach feathers, and stands upright in the antarctic blizzards until incubation is complete. The Antarctic petrel nests on tiny rock nunataks in the inland ice, at least 500 kilometres from the nearest conceivable source of food—a distance it covers repeatedly to feed the growing chicks and across which the fledgelings fly non-stop on their first flight.

The most southerly known land animals are spider-like creatures, the size of a grain of pepper, that live under the edges of loose stones on rocky outcrops far inland in the antarctic ice sheet. They hide from the weak summer sun to avoid dessication, and spend most of their long lives imprisoned in a film of ice on the rock, each in a little cell of gas of its own making. When on rare occasions the temperature and moisture conditions release them, they immediately begin courtship and mating.

Such adaptations fill us with wonder at the endless versatility of life forms. But the very precariousness of their biological existence, and their rapid extinction if conditions change too quickly, make us reflect on the broader consequences for the whole planet if global conditions change rapidly.

There are some common characteristics of polar ecosystems. Food chains are mostly short, as for example from plankton to whales or from lichens to caribou to wolves, and depend more on physical conditions than elsewhere. The species at lower levels—lichens, mosses, algae, and small shrimp-like creatures that live at sea on the underside of the ice—often integrate chemical and nutrient flows over long periods, perhaps decades, while those high in the system— whales, birds, caribou, polar bears, and indigenous humans—tend to be migratory, wide-ranging, or versatile, gathering and integrating dispersed biological energy from large areas and several sources. Such ecosystems are very sensitive to external changes; and because there is little redundancy in the food-chain stages, disturbance of one or two species can have serious implications for the whole system. At the same time, polar ecosystems are remarkably tough, able to survive adverse conditions, and opportunistic in being able to recover when circumstances improve.

## CHANGES IN CLIMATE AND ARCTIC ENVIRONMENT

The possibility that changes in the concentration of carbon dioxide in the atmosphere affect the surface temperature of the Earth was recognized in the nineteenth century; but only in the late 1960s, after observations showed convincingly that the average yearly concentrations were steadily increasing, did the potentially serious implications of the change receive wide recognition (see Chapter 2). It then quickly became apparent that the polar regions, and particularly the Arctic, played a critical role in the processes of rapid climate change.

We have not yet made a satisfactory mathematical simulation model of climate or energy balance in the arctic regions, because of the difficulty in quantifying ocean/sea-ice/atmosphere variabilities, allowing for the effect of multiple cloud layers over reflecting ice and snow surfaces. But a number of international conferences and reviews have resulted in a general consensus between researchers in different countries. For example, the second (1987) international workshop on climate change organized by the United Nations Environment Program, the World Meteorological Organization, and the International Council of Scientific Unions, at Villach, Austria, agreed upon a working scenario of plausible climate change expected in the next half-century. This scenario, conservatively within the predictions of most careful simulation models of global change, became a basis for consideration of policy implications of climate change. Leading investigators from twenty countries examined the likely effects on various regions, including the Arctic. They also considered the changes that might be brought about elsewhere by climate-induced changes in the arctic regions.

The results are necessarily speculative, and the environmental consequences at present are poorly understood. But it is possible on the basis of present knowledge to make some informed guesses in seven important areas: weather and precipitation, planetary heat distribution, sea level, carbon storage, biogeographical zones and habitats, distribution of toxics, and human activities. Here is a very condensed summary of current thinking in some of these areas.

It is important to bear in mind that our concern is based mainly on the anticipated rapidity of change, as well as on the amount or magnitude of changes expected. Climate is always changing, in the Arctic as in other parts of the world; and there have been times when the concentrations of greenhouse gases in the atmosphere were higher than today and about as high as those predicted for the next century. Global temperatures have fluctuated in the past at least as much as the changes forecast by present models. But changes in the past have for the most part taken place slowly, so that biological systems could adjust to different hydrological and meteorological conditions. There have, however, been a few times when significant regional changes of temperature have taken place very rapidly, perhaps in a few decades or centuries, as during some stages of melting of a continental ice sheet. The evidence seems to show that about the same time flora and fauna changed, forests disappeared, and the marine community of northern seas also abruptly changed (see Chapter 4). Yet all these changes, sudden though they were on a geological time-scale, must have been very much slower and more gradual than the climate changes, aided by human actions, that appear to be in store in the next century. All this is why we are apprehensive about the future. And, as in the past, it is in the arctic regions where evidence and effects are most dramatic.

## WEATHER AND PRECIPITATION

The Villach scenario of 1987 assumes an increase of global average surface temperature, due to an increase in greenhouse gases in the atmosphere, of between 1.5 and 4.5 degrees centigrade by about the year 2020. In high northern latitudes, above Latitude 60 degrees North, the average rise in winter temperature is expected to be 2 to 2.4 times more than the global average increase, implying a rise of six or seven degrees centigrade at about Latitude 70 degrees North. Most of the warming will probably be in the continental or coastal areas, and some investigators feel that it will probably temper the coldest periods, rather than be spread throughout the year. The best estimates suggest considerably more polar precipitation in autumn and winter as a result.

Throughout the Canadian Arctic and adjacent seas we might expect, within the next fifty years, summers a degree or two warmer with about the same cloudiness, but more rain and sleety drizzle in the High Arctic semi-desert. In winter, temperatures could be considerably milder, with the prolonged very cold clear weather of the arctic continental interior a thing of the past, but more stormy weather and heavy snowfall on land and at sea. The postulated temperature rises and the increases of open-water sources of moisture could increase five-fold or more the present generally light winter snowfall along the

Arctic mainland coast and in the archipelago. Southern Greenland could expect foggy rainy weather much like present-day Iceland, with more snowfall over most of the rest of the huge island. Warmer winters and increased precipitation can be expected all along the arctic coast of Eurasia, while central Siberia can expect increased winter cloudiness, with drier summers in the sub-Arctic.

The environmental consequences of these postulated weather changes can only be crudely sketched. Generally milder winters and heavier snowfall could have many and various effects: thinner annual sea-ice growth, and more frequent winter storms that would keep the ice moving and broken up; a thicker active layer on top of the permafrost, worsening land stability; increased stream runoff, especially in spring and early summer, and a more vigorous marine ecosystem; higher biological productivity; and more potential human exploitation of arctic seas.

Eventually, of course, continued regional warming would eliminate the sea ice—at what stage we cannot yet calculate. But it is clear that

*Glacier in northern British Columbia. Higher mean temperatures would shrink glaciers and ice sheets, raising the global sea level. (Photo Alan Morgan.)*

the ice would become thinner and disappear from the waters north of Eurasia well before there is much change along the coasts of arctic North America. Thicker and faster-flowing glaciers would produce more icebergs, at least in some areas for some decades, but after a while higher mean temperatures would shrink glaciers and ice sheets.

## PLANETARY HEAT DISTRIBUTION

Warmer arctic winters would be the consequences of more heat flowing from tropics to high latitudes, at first through more vigorous atmospheric circulation, and then through the warmed upper layers of the Gulf Stream and North Atlantic Drift. At its maximum in late spring, arctic sea ice at present covers about 15 million square kilometres (an area about half as big again as the total size of Canada), although from season to season it may fluctuate fifteen to twenty per cent. Reducing the area covered by sea ice would reduce the net annual reflection of solar energy and increase summer heat absorption. Increased winter cloudiness would further blanket heat-loss. The net result would likely moderate the present irregularities in temperature distribution in arctic North America and Eurasia. In time, south-to-north heat transfer would likely reduce with more uniform global warming.

At present cold North Atlantic bottom water from the Greenland and Labrador seas sinks and flows southward down the entire length of the Atlantic, where it then spreads on top of northward-creeping, still colder, antarctic bottom water to fill the lower parts of deep-sea basins in the Pacific and Indian Oceans, and helps to stabilize the world heat-transfer system. Recent studies show that this longer-term energy transfer is sensitive to short-term surface changes. Thus global warming could possibly influence long-term planetary processes.

## SEA LEVEL

While the rise in surface temperature is expected to be quite different in different parts of the planet, a rise in sea level due to melting ice and thermal expansion of ocean waters would within a few months be distributed more or less uniformly throughout the world ocean (see Chapter 3). Rapid melting of the Alaska and Svalbard Glaciers could increase quite quickly the ocean level 10 to 20 centimetres, and the warming postulated in the Villach scenario would cause a rise of 50 to 150 centimetres. Melting of the Greenland Ice Sheet would raise the world sea level about 7.5 metres, but this would take a few centuries rather than decades.

The arctic coastal areas would probably not be much affected, but some local results could be dramatic. The most important would probably be in the deltas of great rivers like the Mackenzie in Canada and the Lena in the USSR, where they would enter the ocean more slowly, leading to quite new sedimentation and delta configuration. This could severely disrupt the rich biosystems of arctic deltas and estuaries, and could change the plumes of silt-laden fresh water that splay along the coasts of the Arctic Ocean from large rivers for distances of 100 kilometres or more and are important to fish and whales.

The Antarctic Ice Sheet, although it has about eight times the area and ten times the volume of the Greenland Ice Sheet, is not at present expected to contribute much to the sea-level rise. It has an enormous temperature inertia, and is still responding to the world temperature transition from a glacial to interglacial climate 15,000 years ago. Its mass is so cold that the postulated climate warming would probably not lead to significant melting, but only to increased moisture and snow accumulation, thickening the ice sheet and withdrawing water from the oceans onto Antarctica. This possibility has been taken into account in calculations about the likely rise in sea level.

## BIOGEOGRAPHICAL ZONES AND HABITATS

The most easily recognized and significant characteristic of the environment of any place, on land or at sea, is the assemblage of plants and animals it supports—the biogeographic zones. These are mainly determined by climate, although local characteristics—such as topography, soils, and sea bottom, and hydrological or ocean circulation patterns—also determine the communities and their habitats within the zones. When the climate changes, zones change. In the temperate latitudes experience in agriculture and forestry shows that the northern limit of common tree species and several grain crops moves northward about 100 kilometres for each one-degree rise in mean annual temperature, provided there is sufficient moisture. But there is no substantial indication that arctic habitats do the same. Boundaries of circumpolar biogeographical zones seem to follow the temperature of the warmest month, not the mean annual temperature, and recent studies of the fluctations of ranges of flowering plants in the High Arctic, and of the boreal treeline, suggest an even more sensitive summer climate control.

If the climate changes are more rapid than all aspects of the ecosystem can adjust to, the system breaks down. For example, the zone of temperature conditions favourable for spruce trees could move northward, by say 200 kilometres, to include presently treeless tundra. If this change took place over three or four centuries, the spruce forest would progressively spread northward. But if the same

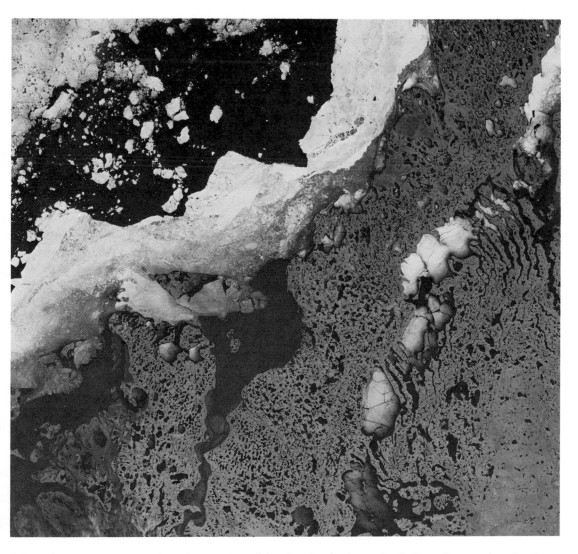

*LANDSAT satellite image of Mackenzie River delta showing ice formation in June. Increased sea levels would change the delta configuration of great Arctic rivers like the Mackenzie and the Lena, severely disrupting the rich biosystems, important to fish and whales, for 100 kilometres or more into the Arctic Ocean. (Courtesy Canada Centre for Remote Sensing; Surveys, Mapping and Remote Sensing Sector, EMR Canada.)*

climate warming took place in a fifty-year period, the chance of seeds developing into healthy spruce trees would be slight, for they would not grow well on soil whose bacteria evolved for tundra, and the necessary bacteria could not possibly migrate northward so quickly. So rapid climate warming may in fact reduce, not expand, the forest— at least until a new equilibrium is established.

What can be said with some assurance is that global warming in Canada would move the temperature ranges favourable for coniferous forests northward onto the generally poorer soils of the Canadian Shield. The area suitable for agriculture would probably be curtailed in

*When climate changes, so does the assemblage of plants and animals. These polar bears are waiting at the ice edge for the water across Hudson Bay to freeze so that they can hunt seals from the ice. (Photo George Hobson, Polar Continental Shelf Project, EMR Canada.)*

the south owing to increasing dryness, and would extend to the north where it would be restricted not by climate, but by the narrowing of the great plains. In Eurasia the same amount of warming, if not disasterously fast, would in contrast likely favour expansion of productive forest across broad areas of the arctic coastal plain and would improve prospects for farming around the White Sea and the lower valleys of the great rivers.

We know even less about habitat change after rapid climate change in the sub-arctic and arctic seas. The present century's northward extension, and then decline, of the West Greenland cod fishery demonstrates how quickly marine habitats can change. Likely warmer winters and more winter snowfall at sea should benefit marine habitats and productivity.

## DISTRIBUTION OF POLLUTANTS AND TOXICS

The Arctic region is significant in world pollution issues not as a source but as a depository, and because of its environmental response to pollution from elsewhere. The Arctic Ocean basin is the ultimate resting-place for long-lasting pollutants from much of the industrialized world. Air and water pollution from eastern North America and Europe is washed into the Atlantic Ocean, and carried northward, by the Gulf Stream and North Atlantic Drift to the Arctic Ocean. There, protected by the ice cover from decompositon brought about by interaction with sunlight and vigorous exchange with atmospheric gases, it remains with little further chemical change, and, as far as we know, accumulates.

Already industrial chemicals and pesticides, manufactured and used in temperate and tropical areas, have been identified in the animals of arctic Canada and north of Norway. It is likely that these chemicals have entered all circumpolar ecosystems. Some have been transported directly in the atmosphere and deposited in the Arctic as polluted snow. Others take the oceanic route, and become successively concentrated in the animals at the top of the arctic marine food-chain—animals that humans then eat. If the patterns of northern atmospheric circulation, the strength and continuity of the Gulf Stream and the North Atlantic Drift, and the capacity of North American, European, and Siberian rivers to flush out chemicals all were to change significantly—not to mention the consequent changes in sea ice-cover, water-mass stratification, and storminess of the Arctic Ocean—the delivery of toxics to the Arctic could be very different.

If long-lasting pollutants from the industrialized world and tropical agriculture are not carried to the Arctic, where then would they go? At present some of our most troublesome wastes are inadvertantly 'swept under the rug' of the Arctic Ocean ice, even if some of it comes back in toxic-laden seals and polar bears, and in migratory birds, born in the North, which carry a tiny amount back south in their bodies and deliver it to the dinner-tables of the lands from whence it came. Natural dumping of toxics in the relatively non-productive Arctic Ocean may be, to some, the ultimate successful demonstration of the NIMBY (not in my back yard) syndrome. To others it represents unacceptable contamination of a region that until recently was little touched by humans, and where there is no possibility of a trade-off between increased productivity and environmental costs. The possibility that human-aided climate warming may alter the present distribution of toxics in the Arctic highlights the essential ethical dimension of climate change and environmental protection.

Old arctic hands remember the unbelievably clear, clean air of early spring in the Far North. Even people who have never been to the North cherish that image. Now every spring, as the sunlight returns to arctic North America, northern Greenland, and the Norwegian arctic islands, there is a brown haze—not as dense as most people in urban areas are accustomed to seeing and breathing, but conspicuous and depressing in a natural and sparsely inhabited land.

Arctic haze is a distinctive polar phenomenon, different from the familiar smoke haze or photochemical smog of industrialized temperate regions. Pollutants originating in Europe and the Western USSR collect in large discrete pancake-shaped masses that hold together in the very cold lower and middle atmosphere of late winter and drift thousands of kilometres across the Arctic to northern North America. In these polluted air masses the concentrations of metals and artificial organic compounds and particulates, mostly carbon, are low, but the dark film deposited on the snow, hastening melting, may have important ecological and hydrological consequences. However, this is one case where warmer winter conditions may have a salutary effect, in inhibiting the travel of polluted air masses such long distances at the edge of the sunlit zone. Arctic haze may thus be one pollution problem that could be diminished if climate changes.

In the higher latitudes, longer-lasting snow-cover means airborne pollutants and acidic compounds accumulate on the snow-surface longer before they are released all at once during melting. This rapid release to soils and waters is known as pollution shock, and typically occurs just at the beginning of the short and vital growing season, which is also the breeding season for wildlife. This may be one of the more insidious human assaults on arctic ecosystems, for it affects germination, growth, and vulnerability to parasites. Arctic-born migratory birds and sea mammals may start life with a pollutant load in their bodies considerably higher than the average concentration of pollutants in the Arctic. The effect of climate change on pollution shock is not clear, but global warming could eventually reduce it.

## HUMAN ACTIVITIES AND GLOBAL CHANGE EFFECTS IN THE ARCTIC

The rapidly changing northern environment, superimposed on the present trends of northern industrial and commercial development, suggests some other intriguing changes and contradictions. Lighter ice conditions may reduce the need for heavy expensive ice-going ships, extend the northern shipping season, and alter domestic and international trade patterns. The Northeast Passage, along the coast of Siberia, can be expected to clear well in advance of the Northwest Passage. There would likely be more oil spills, especially as arctic storminess increases; but with less ice, clean-up would be easier and perhaps the chances of biological recovery greater. A dramatic increase in the number of North Atlantic icebergs could threaten shipping, fishing, and offshore resource extraction off eastern Canada and Greenland. The fish populations on the Barents Shelf and the Bering Shelf may change drastically and erratically, requiring very careful management and international co-operation. Warmer winters, and more snow along the coasts, could increase engineering and construction costs in permafrost areas, but ease water supply and sanitation. Transportation may take on new patterns, particularly in Siberia, where drier conditions in the headwaters may reduce the traffic capacity of the great rivers. Generally warmer winters would decrease the costs of urban-style living, stimulate local small industry, and perhaps make the Arctic more attractive for southerners. They may help northern tourism, particularly if the changes do not damage wildlife and fish. The world's largest national parks, in northeast Greenland and northeast Svalbard, could become world treasures. But climate warming will also make life easier for arctic mosquitoes!

Rapid climate warming does not promise increased biological productivity in most parts of the circumpolar Arctic and sub-Arctic. Indigenous northerners, at present the fastest-growing segments of the populations of every circumpolar country, cannot generally look to predicted global change to increase their sources of local food and livelihood, so important in maintaining their culture and identity. Conspicuous exceptions may be southwest Greenland and the Barents Sea/White Sea area, where productivity may increase. However, a somewhat slower rate of warming than that anticipated now could have quite different and more favourable consequences for northerners in both North America and Eurasia. But a gradual increase of biological productivity could bring the sub-arctic regions more directly into the diverse productive economies of northern nations. In the process, more southerners and southern institutions may move in. It will then be even more difficult to maintain northern indigenous lifestyles and identities.

In the past few decades perhaps the most important influence of the arctic environment on human affairs has been its role in military activities and geo-political strategy. Through the past thousand years, as present political and military systems developed, inaccessibility and the difficulties imposed by the environment have kept the arctic regions a zone of peace, a place of political rivalry without warfare. Circumpolar nations felt protected on their northern flank by a shield of ice and natural inhospitableness that made military threats from that direction impossible. But the advent of the aeroplane, the intercontinental ballistic missile, and the nuclear-powered submarine shattered this shield, making northern nations particularly vulnerable from the North. The response has been an escalation of military activity and investment until, within the last twenty years, arctic regions include the heaviest concentrations of offensive and defensive installations on the planet, and arctic military technology is among the most sophisticated yet devised. To the modern military planner the arctic environment, its sea ice, the noise made by its icebergs, its distinctive magnetic and ionospheric properties, and even its arctic

*Arctic Winter Games at Sachs Harbour on Banks Island in the Canadian Arctic. Climate warming could bring sub-arctic communities into the diverse productive economies of northern nations and make it more difficult to maintain northern indigenous lifestyles and identities. (Photo George Hobson, Polar Continental Shelf Project, EMR Canada.)*

haze—far from being an impassable but passive defence provided by Nature—provide tools and advantages for warfare.

These technological developments, and the geopolitical strategies that lie behind them, have taken place in the arctic environment as it is today. If, through climate warming, the arctic environment were quite rapidly to change its physical characteristics (military strategists and operations tend to ignore anything biological, including resident humans), the effectiveness of present technologies would change considerably. Any opposing forces would have to alter their style of attack, and their defensive and surveillance operations. The use of sea ice as a cover, and the tactics of those who want to find out what is going on under the cover, would certainly change, since lighter sea-ice is often noisier than heavy ice, and the increased open water, and mixing of water layers, would make acoustic surveillance difficult. Similarly, operations on land, in the air, and to some extent in the ionosphere would be affected. Military equipment and investment do not as yet appear to have taken into account the possible effects of global change on the Arctic. Some long-range planners and political leaders, however, show real concern, realizing that the vulnerability of the most advanced arts of war to insidious and simple changes in the arctic environment are evidence of the folly of attempting to settle human quarrels by investing in ever-more sophisticated machinery (see Chapter 12).

Viewed from the south, the region under Arktos, the Great Bear, is indeed unusual and strange in many ways. Studied from within, Arktos is both a sensitive indicator of global change and a teacher of global processes. Traditional Inuit stories attribute to the Great Bear a long memory, intelligence, and the courage to explore new territory and to try something new. Study of the land of the Great Bear can teach us much about what has happened to the planet on which we live, and so revive human memory. It can teach us about ourselves and our fellow inhabitants on this planet, and what we are doing to it, and help us understand what is likely to happen in the near future. Perhaps Arktos can even help us to use our intelligence, and find the courage to develop a new behaviour that will better manage our common home.

## CHAPTER 6

# The Tempering Seas

## LAWRENCE A. MYSAK and CHARLES A. LIN

Containing ninety-seven per cent of the total volume of water on Earth, the oceans play a vital role in global change, the hydrological cycle, and climate change. On the ocean floor the long north-south mountain ranges are the focus of tectonic activity; here continental drift originates, as molten rock from deep inside the earth rises through the ridges and creeps slowly along the ocean floor. Near the ocean surface in middle latitudes, huge wind-driven gyres transport heat poleward from near the Equator. The complex ocean chemistry produces many different elements and compounds that are critical elements in marine life. And finally, the tides, waves, river run-off, and, near the coast, wind-driven upwellings constantly modify the sea level. Less known are the dynamics of wind waves and the major currents that transform the sea surface over the open ocean. We now have advanced satellite technology that can detect even tiny centimetre-small sea-level changes in the open ocean, and give us a picture of the major large-scale current fluctuations.

The oceans are a stabilizing influence on the more rapidly changing atmosphere. They temper it because energy, moisture, and gas are constantly being exchanged, transferred, and converted at the interface of the air and the sea. The global circulation patterns of the seas also act as giant thermal flywheels for the climate system, moving as much heat poleward from the equator as do the winds in the atmosphere. The oceans are obviously a significant component of the climate system, which is made up of the air, water, ice and snow, land, and all the living organisms.

The components of the climate system constantly interact. It is the oceans that initiate and reflect many of these interactions, and this leads to a variety of climatic conditions. For one thing the temperature of the oceans is very stable compared to that of the atmosphere or land surface; while it takes the deep seas a long time to heat up, they also retain their heat a long time, and this can modify the climate of continental coastal regions. El Niño, a short-term warming cycle with widespread and profound effects (see Chapter 2), is one example. The warm Kuroshio Current in the western Pacific south of Japan, where great atmospheric wave-patterns begin, is another, changing the weather faced by North American farmers as far as the mid-west, as well as by coastal fishermen from California to Alaska.

## NATURAL CLIMATE VARIABILITY

We know that climate, the long-term average of weather, varies not only from year to year but from decade to decade. Atmospheric and oceanic records support our observations. Off the west coast of Vancouver Island, for instance, the water in winter is relatively warm for a year or two, then becomes two to three degrees colder for one or two winters, and then turns warm again. The warm years are sometimes related to El Niño events in the tropical Pacific, as in 1940 and 1958. Other warm periods coincide with an unusually strong atmospheric circulation, when powerful winds near the ocean surface have been whipped up by a deep low-pressure system centred over the Aleutian Islands during the winter. These strong surface winds drive warm surface water from the south towards the coast of British Columbia in particular, warming the coastal sea surface. Conversely, when the North Pacific atmospheric circulation is very weak, little warm water flows from the south; cold Gulf of Alaska waters dominate, cooling coastal waters off British Columbia. This correlation is so high that the strength of the winter atmospheric circulation in the North Pacific can be used to predict not only water temperatures, but also interannual sea level changes off the coast.

Sea ice is also important in the heat exchange between atmosphere and underlying ocean. The concentration of sea ice in the Arctic changes from year to year, and this is important in understanding the climate of the northern hemisphere. Both Canadians and Americans tend to assume that their winter weather comes from the north—and they are right. Monthly changes in the ice edge around the Arctic Ocean have been estimated for the last forty years. In the winter, sea ice covers an area of 14 million square kilometres through the Bering and Labrador Seas and across the North Atlantic. By summer it melts back to half the area (Figure 35). The difference is equal to seventy per cent of the area of Canada! The habitat range of fish, seals, and walrus depends on the position of the sea-ice margin each season, and this can make the difference between survival and near-starvation for Inuit who take their living from the Arctic Ocean.

Arctic navigation, resource development, nutrient levels, and fish-

*Ship in sea ice. Sea-ice cover plays an important role in the exchange of energy and moisture between the ocean and the atmosphere. (Dept of Fisheries and Oceans Canada.)*

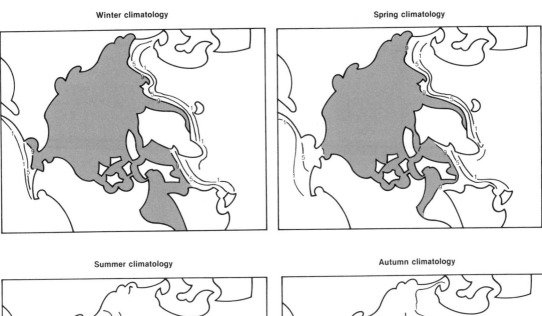

Winter climatology    Spring climatology

Summer climatology    Autumn climatology

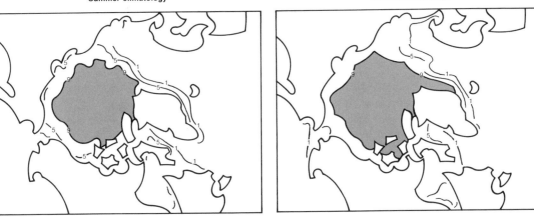

FIGURE 35. *The average seasonal sea-ice extent for winter, spring, summer, and fall in the Arctic, 1953 to 1984. The ice-concentration contours are labelled in tenths (1 = 1/10, or ten per cent of the ocean area along this line is covered by ice; 5 = 50%, etc.).*

eries are also influenced by ice-cover fluctuations, affecting the local as well as the southern economy. In Baffin Bay and Labrador Sea, heavy ice years come roughly every ten to fourteen years. In the cold years of 1972, 1983, and 1984, winter sea ice extended southward almost a third farther than average, and many of the traditional fisheries moved away. Local fishermen lost their fishing grounds and their markets, and their lives were plagued by more cold and wind.

We are not sure why these swings in sea-ice concentration happen; they may be tied to changes in air temperature and atmospheric circulation patterns, which in turn may be related to variabilities in ocean circulation. Whatever the reasons, the shifts are by no means new. The rise and fall of the cod fishery west of Greenland during the late 1800s and early 1900s, and the well-being of the Norwegian and Swedish herring fisheries over the past five centuries, are both ocean/

climate related. Since 1500, Icelandic people have recorded the number of weeks of ice on the north coast of Iceland: roughly every fifty to seventy-five years there was more open water, good for the Norwegian herring fishery. With more ice cover, the fishery packed up and went home (Figure 36).

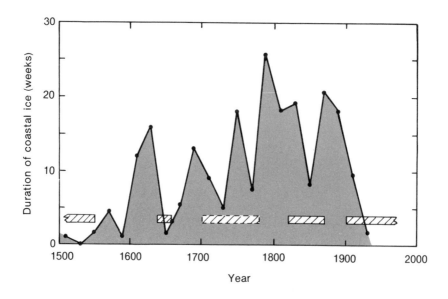

FIGURE 36. *Norwegian herring fishing periods (rectangular boxes), compared with the number of weeks of coastal ice on the north coast of Iceland. (The box marking fishing in the eighteenth century is dashed because of considerable variability.) Severe ice conditions during most of the seventeenth, eighteenth, and nineteenth centuries are consistent with the occurrence of the 'Little Ice Age' in Europe at that time.*

Sometimes the vagaries of the ocean alter fish migration patterns. British Columbia's Fraser River sockeye salmon have a four-year life cycle, beginning and ending in Fraser River tributaries. As juveniles, they travel into the Pacific Ocean through Johnstone Strait, the northern passage between Vancouver Island and the BC mainland. They return to spawn two years later, swimming either south around Vancouver Island, or north as they left. In the late 1950s a substantial fraction returned by the northern route; then most of the fish reverted to the southern passage until 1977, and in 1983 over eighty per cent returned by the northern passage (Figure 37).

Oceanographers then identified two different types of climatic conditions in the northeast Pacific Ocean as causing the shifts. The sea-surface temperature there gradually cooled from 1958 through to

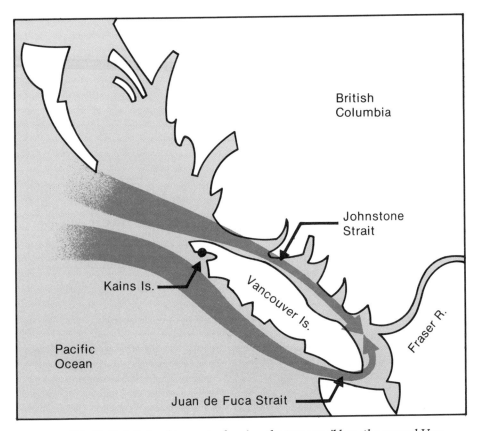

FIGURE 37. *The British Columbia coast, showing the two possible paths around Vancouver Island for sockeye salmon homing to the Fraser River system.*

1976, and during this period the warmer Fraser River discharge was evidently important to the returning sockeye. But with rapid ocean warming—starting in 1976 and persisting to this day—the warm water at Kains Island, near the northern tip of Vancouver Island, drew the fish north (Figure 38).

Predicting the sockeyes' route is important for economic, political, and biological reasons. As many as twenty million Fraser River sockeye return in any one year, and with fish worth tens of dollars each, fishermen want to know where to go. Politically, US-Canada treaty regulations allow American fishermen to fish only in Juan de Fuca Strait, the southern route, and if most fish go further north the regulations may need to be modified. Finally, if one year too many salmon escape the fishnets and make it to the Fraser River, they overcrowd the spawning grounds and not enough fish are reproduced for their next generation.

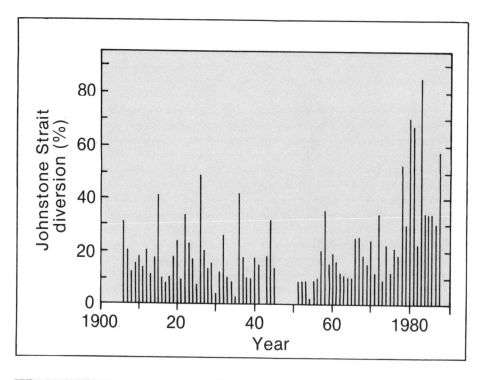

FIGURE 38. *Yearly estimates of the fraction of Fraser River sockeye salmon returning via the northern route, Johnstone Strait. 1953-87 estimates were made by the International Pacific Salmon Fisheries Commission. Pre-1953 estimates are based on historical data.*

*Salmon going upriver to spawn. The sockeye salmon of British Columbia's Fraser River have a four-year life cycle, most of it spent in the Pacific Ocean. (Fisheries and Oceans Canada.)*

## NATURAL SEA LEVEL CHANGES

The sea is never still. Always there are the very regular undulations in coastal sea level due to tides, and the irregular but persistent wind-generated surface waves. But the sea is subject to other, longer-term fluctuations with time scales of several days, years, decades, and longer. As weather systems pass across coastal waters, the resulting fluctuations in sea level can sometimes last more than a week. Mid-latitude cyclones, like the winter Aleutian Low in the North Pacific, can vary from year to year, producing coastal sea level changes of a few centimetres. Hydrological cycle variations, including river runoff, bring other changes over decades or longer. Finally, over these and even longer periods, there is the retreat and advance of the ice sheets, creating fluctuations (Chapter 3).

Looking forward into the next century, we hope, by the analysis of data from monitoring networks, to distinguish between these sea level changes over years and decades and those that may arise from human activity—from greenhouse warming, described later in this chapter.

## CHEMICAL COMPONENTS

Another awesome feature of this element that binds our planet together is that the chemical composition of sea water is remarkably uniform around the globe. The variations in the total dissolved salts such as calcium, sodium chloride, magnesium, and potasium are fairly small; in over ninety-five per cent of the ocean, salinity ranges no more than seven per cent from the mean value of 35 parts per thousand. Interestingly, in some locations—at very high latitudes where the salinity dominates the density changes in the sea—the salt content does vary enough to shift deep ocean circulation. (Temperature is another factor.) When the North Atlantic sea ice freezes, salt is rejected, making the upper part of the water column heavier than the water below. The salty surface water sinks, driving a very deep circulation all the way along the ocean bottom towards the Antarctic. On the other hand, in tropical latitudes the upper waters evaporate, leaving behind more salty water, which is relatively dense and sinks, driving fairly localized deep circulations.

Even changes in the hydrological cycle over the continents make the oceans less salty. During the early 1960s, for instance, there was an unusual amount of runoff in the Canadian North into the Mackenzie River Delta system. Most of this fresh water flowed into the Arctic Ocean and then south with the East Greenland Current through the Fram Passage between Greenland and Spitsbergen. It diluted the surface layer of the ocean, reducing its saltiness so that it froze more readily, resulting in huge extents of sea ice in the Greenland Sea and

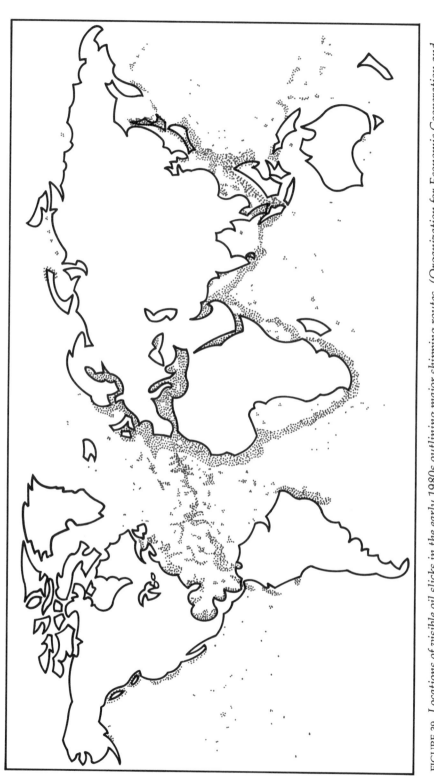

FIGURE 39. *Locations of visible oil slicks in the early 1980s outlining major shipping routes. (Organization for Economic Cooperation and Development,* The State of the Environment 1985.)

later in the Labrador Sea. As we have seen, the extent of sea-ice concentration is important to fisheries and navigation. The same thing could happen if greenhouse warming melts more ice on land, pouring fresh water into the oceans.

The all-important carbon cycle involves chemical changes in oceans. They are one of the two major reservoirs in the global carbon cycle that have a significant control over the amount of excess carbon dioxide released by human activities into the atmosphere; the other is the biomass on land. The oceans have an enormous capacity; their carbon reservoir is more than fifty times that of the atmospheric reservoir.

Three major factors significantly influence $CO_2$ absorption in the oceans: (i) their capacity to take up more carbon dioxide; (ii) the $CO_2$ exchange-rate across the air-sea interface; and (iii) the mixing-rate in the sea. An important aspect of the first factor is the buffering property of seawater. If $CO_2$ in the atmosphere were to increase by ten per cent, the total $CO_2$ concentration in seawater would increase only by about one per cent, depending on temperature, acidity, and total $CO_2$ already dissolved in the seawater. Another aspect is the 'biological pump' process, whereby the phytoplankton in the oceans' surface layers absorb $CO_2$ from the air and then on their death decompose, eventually carrying their burden of $CO_2$ down to the sea bottom.

It is very hard to estimate the global $CO_2$ exchange rate between air and sea, not to mention regional and local variability. More $CO_2$ goes into the atmosphere in equatorial regions because the sea-surface temperatures there are higher. In polar latitudes a colder sea surface draws $CO_2$ in the opposite direction, into the sea.

We also know little about the third factor, the rate of internal mixing. We do know that it happens in the movement of large masses of water, as well as in small-scale turbulence, and that it depends on the difference between $CO_2$ concentration levels at different locations.

## CHANGES CAUSED BY HUMAN ACTIVITY

Changes in currents, winds, atmospheric pressure, ocean temperature, and chemistry have been part of the planet's balancing act from the beginning. Now mankind is altering that balance. One way is by using the seas as a garbage dump and a septic system for sewage, chemical effluents, and even our nuclear waste. As the population increases and land-dumps overflow, it is tempting to turn to the oceans; but we have to wonder if the waste, travelling through complex ocean food chains, will end up back on our dinner tables. (Figure 39.)

Vast as the oceans are, their use as a dump for sewage and toxins poses two types of problems. First, sewage increases nutrient levels, especially near the coast, producing abnormally large algae growth, using up oxygen, and killing fish and other marine populations. Second, heavy metals and other toxic materials dumped into the ocean are trapped in the bottom sediments and remain there for decades. They can accumulate in estuaries, where bottom-layer toxins have closed down many shellfish industries—among others, a once-thriving shrimp industry in the Saguenay River, a tributary of the St Lawrence. The St Lawrence River itself is host to many heavy metals—including mercury from a factory upstream—which have migrated downstream in a toxic sludge that is impossible to catch and remove. People know that the fish are unfit for consumption; but marine animals at the end of the food chain, such as the beluga whales, ingest huge amounts of these highly poisonous toxins. They are ailing, sometimes cancerous, and in 1988 were added to the list of endangered species in Canada.

Exciting more current concern is 'man's great geophysical experiment', the build-up of greenhouse gases such as carbon dioxide in the atmosphere. As discussed in Chapter 2, projecting future warming trends caused by this uncontrolled, unplanned 'experiment' is difficult, depending on population growth, geographical distribution of people, and energy consumption, all hard to predict. However, we also have trouble predicting global uptake and release of $CO_2$ by oceanic and biological processes because of our incomplete understanding of them.

*Clean-up after oil spill in New Brunswick coastal waters. (SSC/Photo Centre/ASC.)*

There is no doubt that the amount of $CO_2$ in the atmosphere is rising. This is shown by continuous measurements since the late 1950s at the Mauna Loa Observatory in Hawaii (Figure 40). And we know the causes: increased burning of fossil fuels, and agricultural and forestry practices.

At present about half the carbon dioxide emissions from fossil-fuel combustion is sucked out of the atmosphere, some going into forests and other vegetation and soils, most into the oceans. Again, the oceans' biological pump performs: surface algae brings $CO_2$ from the atmosphere and through decomposition releases it into the water at greater depths. Along with ocean circulation and turbulent exchange, the carbon dioxide is slowly mixed into the intermediate and deep waters. This mixing takes much longer than it takes the atmosphere to receive and to cycle carbon dioxide.

The huge capacity of the oceans to absorb $CO_2$ means that models of the oceanic carbon cycle can be distorted by even minor misrepresentations of chemical, physical, or biological processes. The estimates of the proportion of carbon dioxide that will stay in the atmosphere, and therefore its influence in warming climate, could therefore be misleading.

### Carbon dioxide concentration Mauna Loa observatory, Hawaii

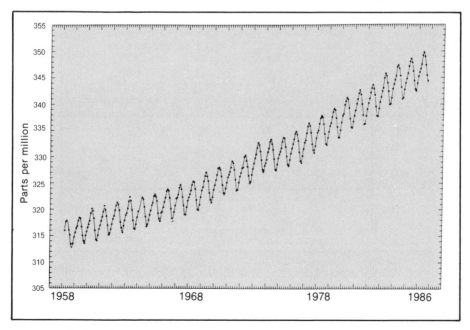

FIGURE 40. *Carbon-dioxide concentration in parts per million (ppm) of dry air, at Mauna Loa Observatory, Hawaii, 1958 to 1986.*

## USING PREDICTIVE MODELS

In order to understand climate variability and predict it for the future, research institutions around the world have developed coupled atmosphere/ocean-climate models. One well-known model has been developed at a National Oceanic and Atmospheric Administration (NOAA) laboratory in Princeton, New Jersey. The Climate Research Group at McGill University has also developed a global ocean-circulation model with a sea-ice component; it is being coupled to the atmospheric general-circulation model of the Canadian Atmospheric Environment Service, to study global and high latitude climatic changes.

The atmospheric component of these coupled models includes solar and terrestial radiation, cyclones and anticyclones, and rain and snowfall. The oceanic component simulates the separate contributions to the ocean circulation of wind action, heat, and salinity.

A numerical experiment with the coupled atmosphere-ocean model

## Model of temperature increase after CO$_2$ quadruples

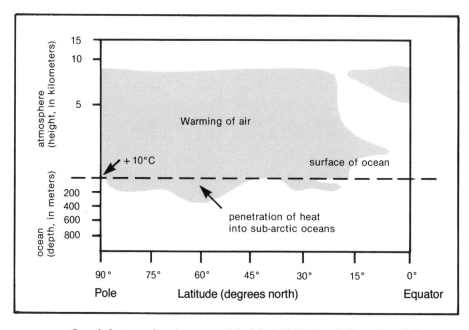

FIGURE 41. *Coupled atmosphere/ocean model of the US National Oceanic and Atmospheric Administration shows the average temperature increase in the atmosphere and ocean one decade after a sudden quadrupling of atmospheric carbon dioxide. The shaded regions show where temperature increases exceed three degrees centigrade, with the maximum of ten degrees centigrade occurring in the polar regions. The excess temperatures shown are averaged over the whole globe.*

starts with the steady equilibrium of atmosphere-ocean climate, a normal $CO_2$ concentration of 300 parts per million. The land/ocean distribution in the northern hemisphere is simplified, for effective modelling of the complex system. $CO_2$ is then increased, say four times, as in Figure 41. The ocean's ability to absorb heat prevents an abrupt reaction in the atmosphere, and the speed of ocean warming depends on the rate at which the excess heat from the atmosphere penetrates. This in turn depends on the large-scale circulation and turbulent mixing processes in the ocean.

The model shows that ten years after the increase of $CO_2$, the largest temperature change is in the lower polar atmosphere, where there is a very warm dome of air over the pole, primarily in winter. In the tropics there is less change, because of strong mixing by tropical weather systems. The ocean is most receptive to heat penetration at about 60 degrees latitude north, that of the southern tips of Greenland and Norway, and of Nunivak in the Bering Sea, because of the subarctic ocean density there.

A decade after the model's enrichment, excess heat has still not penetrated more than 1,000 metres into the deeper ocean. The modelled results of a sudden large increase are easy to read, which is why such a large increase—a quadrupling of $CO_2$—is used. Because the oceans can store so much heat, and take so long to turn over, we

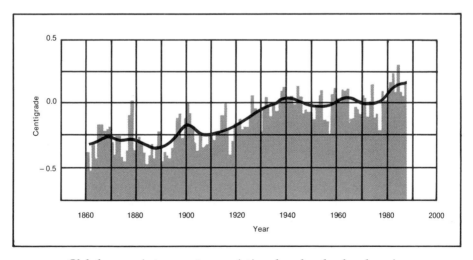

**Global air temperature variations**

FIGURE 42. *Global mean air-temperature variations based on land and marine measurements, shown as departures from a 1950 to 1979 reference period. The available temperature measurements are averaged because they are taken at different unevenly distributed locations around the world and may not be correct in detail.*

can conclude from the model that we would not feel the full impact of even such a large increase for several decades.

What would be the effects of increased $CO_2$ in the atmosphere? Trends today suggest that $CO_2$ will double by the middle of the next century. (A quadrupling, if it occurs, would take place later than that, and is even more speculative.) Model results show that if we double $CO_2$ concentration, we might have a drier and warmer mid-continental North America, Western Europe, and Siberia. (See Figure 58, Chapter 9.) In this scenario, snow melts earlier in these regions, the summer dry season starts earlier because the rainbelt has shifted toward the North Pole, and evaporation is intense in the late spring. Regional cloudiness is decreased, allowing more solar radiation to penetrate and warm the air.

Given a $CO_2$ doubling, the most advanced large atmosphere-ocean climate models recently predicted global warming of 3 to 4 degrees centigrade in fifty to seventy-five years, with about 6 degrees increase in the northern latitudes. With warming of this magnitude, the Greenland ice cap and the alpine glaciers would begin to melt, raising the sea level about 24 centimetres a century. This takes into account only the melting of ice in the northern hemisphere, but the oceans know no boundaries. If we consider the melting of the West Antarctic ice cap, and also the thermal expansion of sea water due to heating, predictions of sea-level rise due to $CO_2$ warming vary from 20 to 140 centimetres or more by the middle of the next century. Large parts of the Canadian and US Atlantic coast as we know it would be innundated; Bangladesh would lose a quarter of its area, and the Maldive Islands in the Indian Ocean would completely disappear!

Because these predictions are based on theoretical models, we must test and check them against observations. Also, we must include in the models the complex chemical and biological interactions among the different climate-system components. Our incomplete understanding of the processes of carbon and heat absorption by the oceans makes it difficult and challenging to model and detect greenhouse warming. Nevertheless, the five warmest years in the last century all occurred in the current decade (Figure 42), and we can say only that these warming signals are certainly consistent with a greenhouse warming. We will probably have to wait several more decades before the full impact, including the release of heat stored in the oceans, is felt.

CHAPTER 7

# Fresh Waters in Cycle

## D.W. SCHINDLER and S.E. BAYLEY

Rivers and lakes are like the blood-vessels of body Earth, carrying nutrients from one part of the planet to another, washing out chemicals and poisons, recycling and returning life-supporting oxygen to the system. Like the bloodstream of a living organism, fresh water can be poisoned, can be affected by the health of the body itself.

Of all water on the planet, ninety-seven per cent is salt, and most of the fresh water is trapped in ice caps. Nevertheless, no ecosystem could function without fresh water. Moreover, the competition between different interests—all demanding use of the rivers, stretching and depleting the resources—is rampant in all developed areas of the world. The demand for energy, transportation, fishing, agriculture, and now for dumping of wastes by great industrial enterprises—nuclear power plants, pulp and paper mills, aluminum companies—multiplies daily. Dams on the Nile in Africa, the Amazon in Brazil, the South Saskatchewan and Churchill Rivers in Canada, all eliminated established farm irrigation or traditional fisheries, and had unpredicted results.

Over half the world's area of fresh water, and fifteen to twenty per cent of its volume, is in Canada. The health of many of the country's regions, as well as basic industries like logging, fishing, agriculture, and manufacturing, all depend on this fresh water. Regional climate relies on it, especially along the great river systems like those of the St Lawrence-Great Lakes and the Mackenzie River.

Most Canadian lakes were formed when glaciers last retreated from the continent 8,000 to 10,000 years ago. Compared to ecosystems in many other regions of the globe, they have had little time to evolve their own unique fauna and flora. As a result, Canadian lakes tend to contain relatively few animal and plant species, and these, because of the lakes' glacial origins, are adapted to cold rather than warm conditions. Many important species of Canadian sportfish, for example, thrive best in regions where air temperatures average between -1 and +2 degrees centigrade (Figure 43).

The young age and glacial origin of the biotic communities in Canadian lakes makes them vulnerable, particularly to significant warming of the sort predicted due to greenhouse gases. Ecosystems with few species tend to be fragile because there is little of what ecologists call redundancy. A northern ecosystem could be compared to a complex business operated by a very few people. If one falls ill, the entire operation is inhibited. On the other hand, ecosystems with a large number

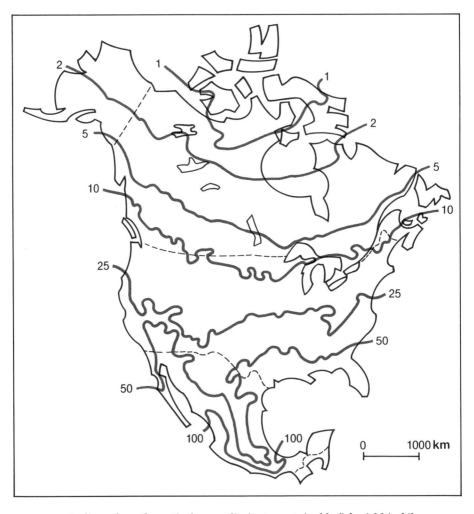

FIGURE 43. *Isolines show theoretical upper limits to sustainable fish yield in kilograms per hectare per year. Most important Canadian sportfish thrive best in regions where air temperatures average between -1 and +2 degrees centigrade.* (Journal of Great Lakes Research.)

of species operate like a large factory: because several species or individuals perform similar tasks, missing elements can usually be provided by others. In Canadian lakes, eliminating one or two species can break links in the food chains that support predatory fishes. For example, three species of crustaceans and two of minnows were eliminated when a small lake in northwestern Ontario was acidified, and the lake trout that depended on them for food starved to death.

*Adult lake trout in Lake 223 of the Experimental Lakes Area, before and after the lake was experimentally acidified. Major food species were exterminated as the lake became more acidified, and the trout starved.*

## RECENT CLIMATE CHANGES

The recent sequence of warm, dry years leads us to ask if climate change is not already happening. Most global models predict maximum warming near the centre of the country in summer (see Figure 58, Chapter 9), so that the earliest certain effects of climate change in Canada would be felt in Manitoba, Saskatchewan, and northwestern Ontario.

There is, however, little support for such scenarios in the long-term record. Central Canada has an unusually long climatic record for a country so young. The Hudson's Bay Company recorded temperatures at many of its trading posts, and some of these records extend well back into the eighteenth century. The dates of ice melt and refreezing—major annual events in a society that relied on water transport—were also carefully recorded. Those records show a long-term trend with a tendency for longer ice-free seasons. A few more years of record should tell us whether these changes really do indicate climatic change.

## CHANGES EXPECTED IN LAKES

Most changes predicted for lakes have been simply deduced from general climate models, and they include features such as water levels for navigation or hydroelectric generation. However, at least some short-term climate changes have occurred during intensive lake studies, so we can make more definite projections. Also, paleoecological studies of lake-sediment fossils help us visualize the probable effects of past climate on lakes.

The physical features of a lake can dictate its response to climate. Small, shallow lakes with large volumes of inflow and outflow tend to respond rapidly to any changes in the supplies of nutrients or water. If a lake is close to the sea, it can be affected by climate-inducing changes in ocean circulation. For example, the algal production of Castle Lake, in the Sierra Nevada of California, was greatly reduced in 1982-3 by El Niño (see Chapter 2), but that of nearby Lake Tahoe, with its enormous depth of 700 metres, was not. Instead, Lake Tahoe's algal production was much higher in years with long, cool springs, when the lake waters mixed more deeply, bringing nutrients trapped in the depths back to the surface, where they could support plant growth.

## THE 1980s—A REHEARSAL OF GREENHOUSE WARMING?

The Experimental Lakes Area (ELA), 200 kilometres north of the Canada-US border in northwestern Ontario, was established by Fisheries and Oceans Canada in 1967. This great natural laboratory, a controlled research preserve, has 46 lakes and watersheds scattered over about 200 square kilometres. Here whole lakes have been used as giant test-tubes to find what acid rain means to fish, animals, and bacteria, what caused the explosive growth of algae in the Great Lakes during the 1960s, and how low-level radioactive elements are distributed in lakes. ELA can also help us predict the likely effects of greenhouse warming on lakes and rivers.

During the twenty years that we have studied the Experimental Lakes Area, there have been several extremely warm, cold, wet, or dry years. At the same time, there has been a strong trend towards warmer weather, causing higher lake temperatures and decreased soil moisture. Our observations confirm that greenhouse warming would have several effects on lakes, including longer ice-free periods, reduced water levels and rates of water renewal, and more forest fires in the watersheds.

A slowdown in the rate of water flowing in and out has a major effect on lake chemistry, altering the lakes' ability to trap incoming chemicals. If drought reduces the amount of water flowing out of the lake, it tends to become more chemically concentrated. This is easiest to visualize for a lake where outflow, and therefore elimination of chemicals, ceases altogether. This actually happens in the closed basin

*Lake 226 in the Experimental Lakes Area was fertilized with nitrogen and carbon in the near basin and with phosphorus, nitrogen, and carbon in the far basin to study the effects of phosphorous on eutrophication.*

lakes of southern Saskatchewan and southwestern US, causing chemicals to reach concentrations so high that they precipitate, or solidify, from solutions. Even sodium chloride, although it is the most soluble of common natural salts, is sometimes precipitated.

Water renewal is so important that it is a major component of models predicting the effects of both eutrophication (nutrient enrichment) and acid rain on lakes. As water-renewal models would predict, chemical concentrations in lakes at ELA have increased during dry periods, especially during successions of dry years such as in the 1980s (Figure 44). The most significant chemicals concentrated are nutrients like phosphorous and nitrogen, which stimulate algal growth. The effects are similar to those caused in lakes by other human activities, such as discharge of sewage and fertilizer, which cause algae blooms to flourish and rot, threatening the diversity of aquatic communities with lack of oxygen. Reduced soil moisture and warmer temperatures in a relatively arid forested area like ELA may also increase forest fires. Area forests have burned over about every 100 years for the past one or two thousand years, because of a combination of dry weather, lightning strikes, and insect or disease outbreaks. During the driest periods the lake watersheds have burned much more frequently, in extreme cases every decade or two. The dry, hot late-1970s and early-1980s saw repeated forest fires; some watersheds burned twice in less than a decade, so the landscape looked something like it did shortly after glaciation. Fires increase the run-off of nutrients and other chemicals from watersheds to lakes—at least until forests are established again.

Other studies give us a preview of the effects of climatic warming in lakes. Cooling water from thermal or nuclear power plants, for instance, dumped into lakes, causes a longer ice-free season and faster nutrient recycling. The increased algal productivity is usually passed up the food chain, from one nutritional level to another. Warm-water inputs to lake surfaces also prevent the normal seasonal turnover of the layers of water, each with different degrees of heat. With higher productivity, this may cause more intense and prolonged oxygen depletion in cold bottom waters, a threat to cold-water organisms.

It seems at first glance that warming may have an overall positive effect on lakes, but qualitative changes in the growing organisms make this doubtful. Many northern temperate lake organisms are what scientists know as cold stenotherms; that is, they are unable to tolerate warm conditions. We believe that most of these species, called glacio-marine relicts, originated in arctic oceans, became trapped, and were pushed south into freshwaters by advancing glaciers. When glaciers retreated, they were left in residual lakes. After thousands of years of adaptation to cold, many glacio-marine species cannot

*Columns of calcium carbonate loom above California's Mono Lake. Such deposits are formed when evaporation causes lakes to become chemically supersaturated, causing precipitation of calcium carbonate and other salts. (Photo Alan Morgan.)*

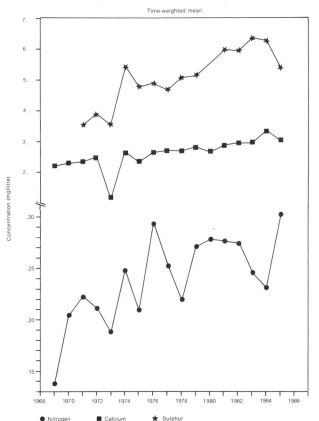

**Change in elements – lake 239 Epilimnion**

Time-weighted mean

Concentration (mg/litre)

● Nitrogen  ■ Calcium  ★ Sulphur

FIGURE 44. *Changes in the concentrations of calcium, sulphur, and nitrogen in ELA Lake 239, 1970-87, a period hotter and drier than normal. The increases are due to slower water-renewal rates for lakes and more chemical run-off from areas burned by forest fires.*

tolerate temperatures warmer than 15 degrees centigrade. The dozens, if not hundreds, of such species include lake trout, lake whitefish, opossum shrimp, and many others important in northern aquatic food webs. Further south these species cannot survive in near-surface waters during midsummer and retreat to cool lake bottoms. In lakes that are shallow and warm, or are eutrophic enough to have significant oxygen deficits in deep waters, they die. The deforestation of watersheds by fire increases exposure of lakes to the wind, causing warmer water to deepen, further reducing the habitat for the cold-loving species. At cooler, more northerly latitudes, cold stenotherms inhabit surface waters year round, in shallow as well as deep lakes.

Desirable warm-water species could of course thrive in the same habitat. For example, smallmouth bass are currently restricted to extreme southern Canada, and many northern fishermen would welcome this species. However, the movement of fish from cold to warm water will not occur automatically. Most higher organisms in freshwaters have relatively poor dispersal powers—lake trout and other species, introduced during glaciation, have not changed their distribution patterns since that time. If productive fisheries are to continue in these lakes, it may be necessary to stock them artificially with species that tolerate warmer conditions.

## IMPLICATIONS FOR MAJOR CANADIAN FISHERIES

Temperature increase in Canadian lakes would not be a trivial problem. Many of the largest fisheries in central Canada occur in shallow lakes at the southern limit of distribution for lake whitefish and lake trout in surface waters—for example in Lakes Winnipeg, Manitoba, Winnipegosis, and Southern Indian Lake. Together, these fisheries are worth $21 million annually. But in a time of prolonged drought, Lakes Manitoba and Winnipegosis could dry up completely, as they did several times between 6,000 and 4,500 years ago.

In more northerly latitudes still other problems may be caused by climatic warming. From an average latitude of about 53 degrees North, about that of northern Lake Winnipeg or southern James Bay, the banks of many lakes and rivers are stabilized by permafrost. If this begins to melt, enormous bank slumping results, sweeping many tons of fine silts and clays into fresh waters, burying bottom organisms and blocking the light available for photosynthesis.

## RIVERS AND STREAMS

Most of Canada's electrical power is generated from rivers, and any significant decrease in flow—possible if the climate becomes hotter

*Bank slumping on Southern Indian Lake in northern Manitoba, because of the melting of permafrost. (From Newbury and McCullogh,* Journal of Fisheries Research Board, *Canada, 1984.)*

and drier—would lower potential power production. Replacing hydroelectrical output, lost through climatic change, with nuclear and fossil fuel would cost the province of Ontario alone an estimated 64.6 million dollars annually, according to the International Joint Commission. The large hydroelectric installations in northern and central Canada would probably be even harder hit because of the expected climate warming, but so far there have been no cost estimates for their replacement. On the other hand, of course, increased power generation using fossil fuels would cause increased release of carbon dioxide to the atmosphere, amplifying the greenhouse effect.

Navigation, too, could be affected by climate warming. On the Great Lakes, low water levels and a longer open water season are expected; the difficulties of the first outweigh the benefits of the second, with a thirty per cent predicted annual increase in shipping costs. But the port of Churchill on Hudson Bay should offer a longer ice-free season, perhaps increasing shipping via that route.

Changes in river hydraulics and levels alter habitats for particular species. Although there are few detailed studies, obviously lower water levels must affect fish like salmon and char that go upriver to spawn. Seasonal flow patterns will also probably change with decreased average river-flow rates, owing to the increased diversion and damming of rivers for power, irrigation, and other human activities. If seasonal water-level fluctuations cease, many diverse wetland and deltaic communities, which rely on dynamic seasonal flow patterns for replenishment, will be gradually eliminated.

## RIVER DIVERSIONS

In intense, prolonged droughts people in parched cities and farmlands may look north, eyeing the remote lakes and rivers and demanding their diversion south. Such diversions are already under way in the Soviet Union, where seventy-five per cent of the population live in the arid southwest, but eighty per cent of the drainage is by rivers that flow to the arctic coast. Damming and diversion of western and central USSR rivers have brought millions of hectares of marginal lands into cultivation, improving navigation and increasing hydroelectric power. These massive engineering schemes were preceded by hundreds of studies by over 120 institutions over several decades. Nevertheless, they have radically changed the rivers' chemistry, temperature, and flow, damaging fisheries and curtailing the planned benefits. The most grandiose plan—to divert water south from Arctic-bound rivers—has been shelved.

Monumental diversion schemes have also been planned in North America, and repeatedly dismissed as uneconomical. The drought of the 1980s, and the possibility that Canada might compromise future water resources by inadequately protecting them in trade agreements with the US, have revived concern.

Two major schemes have been proposed for North America: the GRAND project in the East, and NAWAPA in the West. GRAND would dam James Bay, slowly converting it into a huge freshwater lake, backing water overland into the Great Lakes, from which it could be easily routed south. However, the potential problems are enormous, ranging from regional climate changes to the invasion of exotic species into the Great Lakes, where the St Lawrence Seaway has already destroyed the natural food chain by introducing sea lamprey, rainbow smelt, and alewife, which destroy the 'natural' food-chain. Of course the marine fauna of James Bay would be almost totally exterminated, as they have been in the similar but smaller conversion

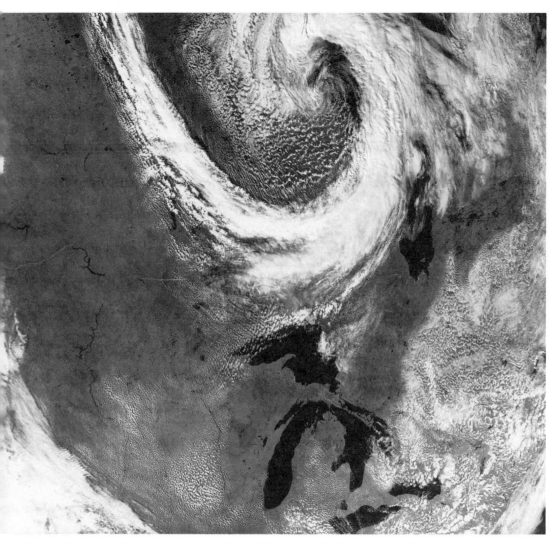

The Great Lakes are the largest grouping of inland freshwater bodies in the world. The proposed diversion of water from James Bay into the Great Lakes and further south could have significant climatic and environmental impacts. (Courtesy Canada Centre for Remote Sensing; Surveys, Mapping and Remote Sensing Sector, EMR Canada.)

of the Dutch Zuider Zee from salt to freshwater for flood control.

NAWAPA, an even more grandiose scheme, would divert water from northern rivers in western Canada, including the Mackenzie and the Yukon, flood the Rocky Mountain Trench and transform it into a vast canal carrying water to the arid southwest US (Figure 45). The north-flowing Mackenzie River now functions as a vast heat pump, warming northern regions with warm water from southern parts of Canada and producing a considerably more moderate climate than in other areas at the same latitude. Benefits include earlier ice-melt dates

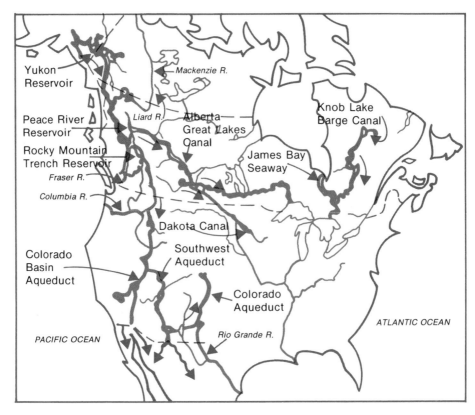

FIGURE 45. *Proposed major river diversions in North America. (From Richard C. Bocking in* Canadian Aquatic Resources.)

and later freeze-up in the Mackenzie and in nearby lakes. Obviously diversion of the Mackenzie's water would remove this heat source. The navigation season would shorten, seasonal floods that replenish Mackenzie Delta biological communities would weaken, and the Mackenzie Valley would lose its privileged, relatively warm climate. The drastic effects on northern-delta ecosystems were illustrated by the first major diversion of water from the Mackenzie system, when the W.A.C. Bennett Dam was closed in 1967. Within two years adverse effects were noticed in Alberta's Peace-Athabasca delta area, which had been replenished by annual spring floods from the Peace River before it was dammed. Invasion of exotic species could be even more devastating than in the GRAND project, for the much smaller number of species in western ecosystems makes them far more likely to be eliminated or displaced. Flooding the trench would also eliminate one of the most magnificent and massive landscapes in Canada, home of both indigenous and white people.

## WETLANDS

If lakes and rivers are the blood-vessels of ecosystems, wetlands are the kidneys, removing nitrogen and sulphur from the atmosphere and recharging ground water, lakes, and streams.

Wetlands are known by a number of different names across the world. In Canada we make a distinction among marshes, swamps, bogs, fens, and shallow water. In other parts of the Earth they may be called mires or moors, or by a host of names peculiar to the world's languages. Whatever they are called, they are characterized by land that is saturated with water long enough to produce plants and soils characteristic of wet environments. Wetlands—particularly the huge tracts of peat-bogs between 50 degrees and 70 degrees in the northern hemisphere—contain the second-largest pool of carbon in the world, exceeded only by the tropical forests. In Canada wetlands take up more surface—fourteen per cent of the total area—than do lakes. They are mostly in Manitoba, Ontario, and the Northwest Territories.

Some wetlands, like saline sloughs in the North American plains,

*Valuable wetlands at Point Pelee, Ontario. (Photo E.W. Manning, Environment Canada.)*

are discharge areas, where ground water moves toward the surface. Wetlands are breeding-grounds for birds, especially water fowl, fur-bearing animals, a variety of small mammals, and some fish species, supporting many rare and threatened animal and plant species. Greenhouse warming will probably cause substantial changes in wetland ecosystems, and in the roles that wetlands play in the global bio-geo-chemical cycles of carbon and other elements. Some tend to think of wetlands as wastelands, convertible to more productive uses like agriculture, highways, housing, and industry. Only now are we recognizing their value as moderators of watersheds, wildlife habitat, recreational areas, and as a sanctuary of the planet's threatened biodiversity.

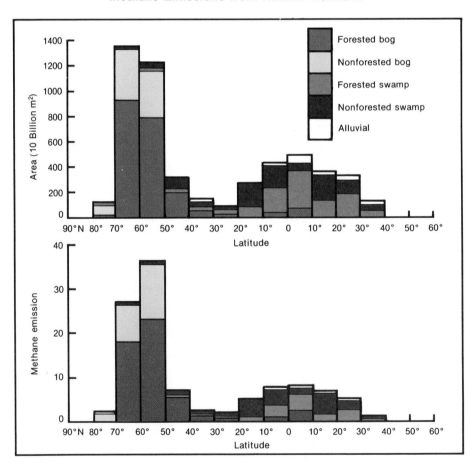

FIGURE 46. *Latitudinal distribution of wetland areas and annual methane emissions. (From Mathews and Fung,* Global Biogeochemical Cycles, *1987.)*

Areas such as the peat-bogs of the Hudson Bay Lowlands are the largest single contributor of methane, one of the greenhouse gases, to our atmosphere (Figure 46). But it is unlikely that these northern wetlands are responsible for increases in the amount of methane in the atmosphere in recent years. However, changes in temperature or water levels could alter the amounts of methane, carbon dioxide, and other gases released in the future, making it all the more important for us to understand the controlling environmental factors. If the climate gets wetter, for instance, wetlands will discharge more methane. If it gets drier, oxidation of organic matter will produce more carbon dioxide, another greenhouse gas.

North American prairie wetlands are primarily marshes growing a variety of grass-like plants: rushes, cattails, reeds, sedge, grasses, and many broadleaved moisture-loving herbs. A warmer global climate would change these wetlands radically, and they are the most important habitat for much of the continent's waterfowl. The Prairies suffered droughts in the 1930s, and again in the late 1950s, and each time lost some of its wet potholes. In the 1980s we witnessed

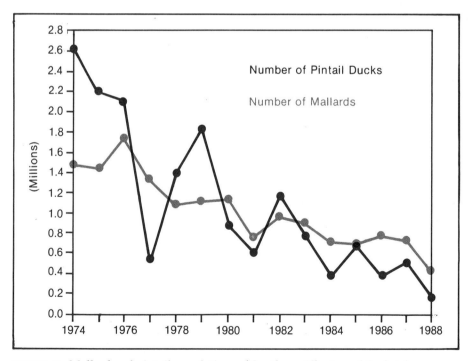

**Duck Population 1974-1988**
**Southern Alberta and Saskatchewan**

FIGURE 47. *Mallard and pintail populations of Southern Alberta and Saskatchewan, 1974-88. (Environment Canada.)*

another drought during which the number of marshes has fallen from 6 million to 1 million. The result has been devastation for the prairie-duck population (Figure 47).

Ever since the Prairies were broken for agriculture, wetlands have been drained for cultivation. A drained pothole and a naturally dried-up pothole are the same; the vegetation that supports waterfowl is gone. Recent dry conditions have made it easier to drain prairie marshes, converting them to agricultural use and losing them forever.

Drainage and the drying of other wetlands cause changes in plant composition. Occasional high-water periods are essential to prevent woody plants from invading shoreline wetlands. If water levels fall, or are stabilized by dams, wet thickets—with less biological diversity and productivity—replace the rich shoreline marshes.

Another consequence of the drying of wetlands is that the elements of water become more chemically concentrated. In some areas this will cause eutrophication, changing wetlands as well as lakes. Many rare plant species grow only in wetlands without much vegetation, and these areas are becoming rare as algal growth increases.

In coastal areas, global warming will cause a rise in sea level, flooding, and in many cases the destruction of productive fresh-water marsh communities that exist in the upper reaches of estuaries. We have already seen this in the marshes of the Chesapeake Bay and Louisiana. Some freshwater marshes will be converted into brackish estuarine marshes.

The effects of climate on wetlands are aggravated by industrial activity. Wetlands are unique in the way they store and release acids. Under normal conditions bog and fen vegetation removes sulphate from rain and snow and stores it in plant tissue, neutralizing acid rain. In dry weather more sulphur is converted to acid, and aquatic organisms can be harmed when the rain flushes the reoxidized sulphur from marshes into lakes and streams. If the Earth's atmosphere continues to warm, reducing wetlands even further, the amount of sulphur-fixing vegetation would be diminished. Thus acid precipitation could course more freely into our lakes and streams, wreaking havoc on aquatic life.

*Ever since the North American plains were broken for agriculture, wetlands have been drained for cultivation. By 1981, 76 per cent of the original wetland areas surrounding major Canadian prairie cities had been converted to agriculture, as here in the Fraser River Delta of British Columbia. (Environment Canada.)*

## ACID RAIN

Climate is only one of several changes expected following the combination of insults known as global change. Acid rain was first recognized in the British midlands in the mid-1800s, but the continental scale of its damage became evident only in the middle of this century. Acid rain is generated by a combination of atmospheric pollution, notably from sulphur oxides from the burning of fossil fuels in industry and from home heating and cooling, and from the smelting of ores with high sulphur content. The damage it can do to a specific area depends on the ability of the receiving terrain to neutralize acid, and this is largely determined by the nature of soils and bedrock (Figure 48). Widespread acid rain, discovered in northern Europe and Eastern North America since the 1960s, now threatens western North America, Japan, China, the Soviet Union, and South America, and has already caused acidification of many aquatic ecosystems in those areas. Most adult sport fish seem to be able to tolerate fairly high levels of acid, but young fish and organisms lower in the aquatic food web suffer. When they die, large predatory fish can starve before they succumb directly to acid rain. Reducing sulphur-dioxide emissions—as in the Sudbury area in central Canada, where a combination of nickel-smelter closures and other controls has cut sulphur pollution to about one-third of the rate in the early 1970s—allows rapid chemical recovery of lakes. However, suitable conditions for the original species in many lakes may require still further reductions, particularly in the USA, which is where much of the acid rain falling in eastern Canada originates.

There will be a number of synergistic effects of these assaults on water resources. For example, the drought expected to accompany climatic warming in much of Canada would cause reoxidation to sulphuric acid of reduced sulphur compounds that have been stored in wetland vegetation and watershed soils as the result of acid rain. Even small precipitation events during such droughts could release acid pulses to lakes and streams. Because the expected forest fires enhance such reoxidation, more sulphuric and nitric acids would enter the water. Drought, caused by climate change, and atrophy of tree roots in acidified soils, where toxic aluminum has been dissolved from minerals by acid rain, would add to the stress on trees. And there are many more examples of the results of stress on water resources.

In addition to these snowballing effects, many human reactions can interact to worsen the consequences of global change; river diversions are just one illustration. We saw another bad situation made worse in the response to the 1988 drought. In southwestern Manitoba, farmers used local well water on young crops, trying to sustain them in the hope of rain. As a result, many municipal wells went dry, lowering water tables even further. Much of the water evaporated before it had any effect on crops. Also, in dry years farmers tend to till dried-out wetlands, reducing the effectiveness of these ecosystems as ground-water recharge areas and waterfowl breeding grounds in more normal weather.

Many of these misuses can be regulated. Obviously if global change occurs as scientists predict, we will have to change the way we use water. It will certainly be less plentiful, and unless major studies of consequences for the habitats of fishes, waterfowl, and fur-bearers begin now, our harvests from these resources will be diminished.

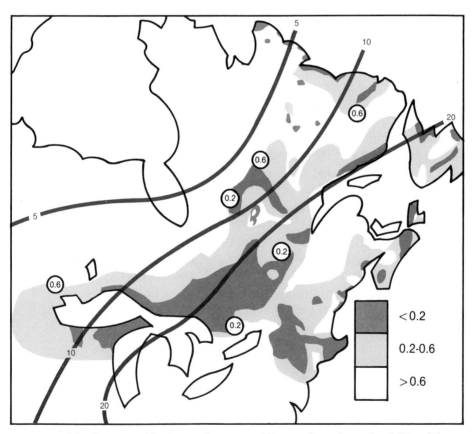

FIGURE 48. *The damage acid rain can do to a specific area depends on the ability of the terrain to neutralize acid; this is largely determined by the nature of soils and bedrock, as indicated by the average ratio of alkalinity to calcium and magnesium in freshwater lakes. The northern Great Lakes region of North America is extremely acid sensitive. This is an area with high acid deposition because of industrial pollution. Dark areas show high sensitivity; the ratio in pristine lakes usually ranges from 0.6 to 1.1. Heavy lines numbered 5, 10, and 20 indicate sulphate deposition in kilograms per hectare per year, and were added by the author.*

# CHAPTER 8

# Forests:
# Barometers of Environment
# and Economy

### J.S. MAINI

Strong ties have bound the destiny of men and trees since the dawn of human history. Trees and forests have been woven intimately, by peoples of every continent, into culture, religion, and mythology, as well as exerting a profound influence on survival and economic development. About 10,000 years ago, before the dawn of agriculture, rich forests and open woodlands covered an estimated 6.2 billion hectares of Earth's land surface. Now Earth's total forest land has shrunk to about 4.3 billion hectares—a loss of nearly one third. France, once eighty per cent forested, by 1789 had trees covering only fourteen per cent of its land. In continental United States, forests covered 385 million hectares in 1630, but by 1920 this tree-cover was two-thirds gone. In Canada, 14.6 million hectares of forestland have been cleared and converted to farmland since the arrival of Europeans—more than the size of the three Maritime Provinces together.

## FORESTS AND ECOLOGY

Forests are the heart and lungs of the world. Most river systems originate in forestland, or from lands that were once forested, and most watersheds are forested or have been under tree-cover in the past. The drainage systems of many of the world's great rivers—among them the Amazon, the Ob-Irtysh in the western Soviet Union, the Congo in Africa, and the Columbia in the US—are forested. In Canada, forested slopes of the Rocky Mountains, as well as the boreal forest, give birth to many rivers. Forest-cover prevents soil erosion, maintains high water quality, regulates seasonal waterflow, and consequently serves as the bloodstream of agriculture, industry, and

*Arrowhead Lake, north of Slocan, BC. Forested mountain slopes, in Canada and else-where, give birth to our many rivers. (Photo Forestry Canada.)*

human communities. Indiscriminate forest removal on watersheds has had disastrous results, as we have seen with deforestation of the Himalayas and many other mountainous regions: this was followed by loss of soil and land productivity, sedimentation of irrigation systems and turbines, and flooding in the plains.

Trees are referred to by some as the 'lungs of the world', because they 'inhale' carbon dioxide and 'exhale' oxygen. Through the process of photosynthesis, this carbon is 'fixed' and becomes a part of forest biomass (wood, bark, and leaves) and constitutes an important part of the global carbon reservoir. The oxygen released, however, has a minor impact on atmospheric oxygen concentration. Forests are the world's greatest engine for storing the Sun's energy, which they do by converting carbon dioxide and water into living tissue. In one day a

*Deforestation of mountain slopes leads to erosion of soil and flooding in the plains below, as in Nepal. (Phillip Tresch, CIDA photo.)*

single deciduous tree can pump and transpire as much as 20,000 litres of water, taking up carbon dioxide at the same time. By doing this, trees serve as an important reservoir for carbon dioxide, helping to regulate the global carbon dioxide balance and the carbon cycle. Collectively they are estimated to contain two thousand billion tonnes of carbon—400 times the amount released each year by burning fossil fuels. Deforestation, and the burning of wood as fuel, diminish the carbon reservoirs and increase carbon dioxide emissions to the atmosphere. Deforested land that is not regenerated back to tree-cover removes the cut area from the process of storing carbon. Destruction of forests over past centuries has contributed about a fifth of the total increase in atmospheric carbon dioxide concentration: 90 to 180 billion tons since 1860, compared with 150 to 190 billion tons from the burning of fossil fuels.

The assault on our atmosphere by the over-abundance of carbon dioxide being released from the Earth is continuing and complicated. Carbon dioxide emissions from the land include vegetation and forests, and from fossil-fuel burning in industry, home heating and cooling, and automobiles. At the same time $CO_2$ is absorbed and 'fixed' by oceans and vegetation. Our concern is the increasing net amount of carbon dioxide that is being added to the global atmosphere every year, along with other greenhouse gases, which are associated with the anticipated climate change. Deforestation—plus decomposition of litter and other forest organic matter—has been a major contributor of carbon dioxide in the past, responsible for up to half of the $CO_2$ emissions, but today eighty per cent are from fossil fuels.

Forests also filter particulate pollutants from the air, and improve local and regional climate, both tropical and temperate. Trees themselves are also subjected to a wide range of airborne pollutants, including acid rain, which reduces tree growth and makes them more susceptible to insects and diseases, as well as injuring other forest flora and fauna, both above and below ground.

Forests have another vital role that affects our survival: particularly in the tropics, they are a major source of global genetic diversity and centres of species evolution. Without the green forest mantle, much animal life as we know it would never have evolved; its diversity ensures that life will continue, that new species can evolve under a constantly changing environment. Although tropical forests cover only seven per cent of the world's land area, they contain over half the known plant and animal species, including eighty per cent of the insects and ninety per cent of the primates. Consequently large-scale forest clearance can have a devastating impact on global genetic diversity, and on gene pools of individual species.

## REGENERATION GAP

The difference between forestland harvested and land regenerated by trees—the 'regeneration gap'—continues to increase every year. In the tropical regions in the early 1980s, about 11.3 million hectares were

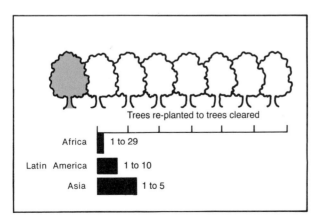

FIGURE 49. *Disappearing forests. There is a regeneration gap between trees cut and trees re-planted. It is particularly severe in the Third World. (Worldwatch Institute, Washington DC, 1985.)*

*A nursery of eucalyptus trees in Kenya. The gap in area world-wide between forest-land harvested and land regenerated by trees continues to increase every year. (Photo N. McKee, IDRC.)*

cleared annually and only ten per cent was planted; in Africa, the ratio was 29 to 1, in Asia 5 to 1 (Figure 49). India, over half forested in the early part of this century, was down to 14.1 per cent in the early 1980s.

## FORESTS AND THE ECONOMY

We are only now learning to value the ecological role of forests, but their role in economic development has been recognized and evolving over thousands of years. Nearly half the world's main crops originated in tropical forests. Forest products, as well as forest processes, are used as a basic life-support system in a subsistence economy. In emerging market economies, forest harvests generate the necessary capital for economic development. Developed economies—such as most Western countries, including Canada—located in the circum-polar belt of boreal forests, have now made major capital investments in their forest sector. Any shift in the quality and quantity of forests—either through indiscriminate deforestation, decline due to air pollution, or climate change—would have far-reaching economic as well as environmental and social consequences.

Collectively, forests contain an immense bank of the chemicals we use pharmaceutically. While a quarter of the drug prescriptions in the United States use active ingredients from plants, only one per cent of

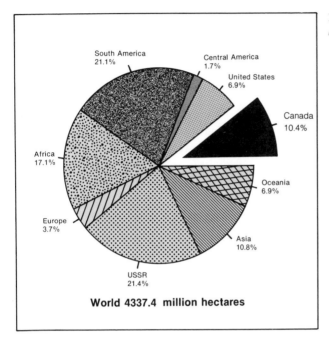

FIGURE 50. *World forest areas by regions.*

South America
21.1%

Central America
1.7%

United States
6.9%

Canada
10.4%

Oceania
6.9%

Asia
10.8%

USSR
21.4%

Europe
3.7%

Africa
17.1%

**World 4337.4 million hectares**

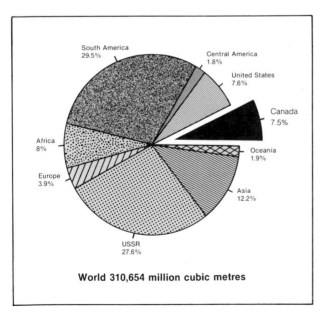

FIGURE 51. *World forest growing stock.*

South America
29.5%

Central America
1.8%

United States
7.6%

Canada
7.5%

Oceania
1.9%

Asia
12.2%

USSR
27.6%

Europe
3.9%

Africa
8%

**World 310,654 million cubic metres**

the species has been investigated; researchers estimate that 1,400 forest plants may have anti-cancer potential.

Of the total global forestland of 4.3 billion hectares, nearly sixty per cent is in the USSR, South America, and Africa. Only a fifth is in North America, half in Canada (Figure 50). Area is only one indication of the forestland's ecological and economic significance; biomass and species diversity are also important. Whereas forests in Canada, the USSR, and South America cover 10.5, 21.4 and 21.1 per cent respectively of the world's forestland, they produce 6.5, 28.3 and 30.1 per cent of the total volume of wood, showing the comparatively low biological productivity of Canada's forests, in northern latitudes and temperate climate (Figure 51). The circumpolar boreal forest belt constitutes a significant component of the world's carbon reservoir on land, which is important to the global climate.

As Canadians are stewards of about a tenth of global forestland, its ecologically sustainable development is both a national and a global responsibility. To fulfil it, we must understand three principal components: the nature of atmospheric change, the potential response of forests to anticipated change, and human activities that have contributed to change (Figure 52). All three subsystems interact, constantly evolving.

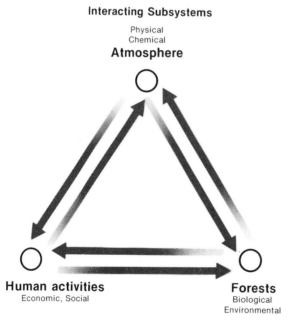

**Interacting Subsystems**

Physical
Chemical
**Atmosphere**

**Human activities**
Economic, Social

**Forests**
Biological
Environmental

FIGURE 52. *Three principal interacting subsystems, which are constantly changing, are involved in the stewardship of the world's forests.*

## ATMOSPHERIC CHANGE

How complex is the ecological interaction between atmosphere, trees, and forests! The growth of individual trees, the zonal distribution of different forest types, tropical or temperate, and latitudinal as well as altitudinal tree-lines are all influenced by climate. The global climate has been relatively stable over the last 10,000 years, with the warmest time only one degree centigrade warmer than today. But over the last century the globe has been very slowly warming, increasing its average temperature about 0.5°C in all that time. However, atmospheric change is not new. In the ancient past, geologic events such as continental drift, glaciation, and volcanic activity all altered the atmosphere, and therefore plant and animal life. What is unique about the present phase of atmospheric change is the human factor. Human activities have now become the dominant change factors, proliferating greenhouse gases during the past several decades, warming the earth's surface and the lower atmosphere, and potentially changing the climate (see Chapter 2). The northern hemisphere, where ninety-five per cent of carbon dioxide emissions originate, is also the principal recipient of feedback from the system. The impact of atmospheric change is likely to be more pronounced in the middle and northern latitudes. Temperature increases will likely be higher in the high latitudes, and during the winter. We expect the growth and survival of individual tree species to be significantly affected over vast areas.

Soil moisture will likely be reduced in summer over extensive mid-continental regions of both North America and Eurasia in middle and high latitudes. Over northern Canada and northern Siberia, snow is likely to melt earlier, reducing soil moisture available for plant growth (see Chapter 9).

## IMPACT ON TREES AND FORESTS

Vegetation types depend on climate, especially temperature and precipitation during the growing season. Forest types mapped across the world show a band of boreal forest—tall conifer formations with some deciduous trees—stretching across the top of North America from Alaska to Labrador, and across Eurasia from Scandinavia to Siberia. This band accounts for sixteen per cent of forest cover. South of that is the temperate forest, a more diverse mixture of evergreen and deciduous hardwoods, covering Great Britain, much of Europe, the central Soviet Union, and northern China. In North America it extends down the Pacific coast and over the eastern half of the US. Tropical evergreen forest, the most luxuriant and diverse of all, ranges

*Boreal forests account for about 16 per cent of the world's forest cover. This one is near Slave Lake in northern Alberta. (Photo Forestry Canada.)*

*A temperate forest in southwestern Quebec. Such a diverse mixture of evergreen and deciduous hardwoods once covered Great Britain, much of Europe, central Soviet Union, and northern China, as well as the eastern half of the US. (Photo Forestry Canada.)*

around the Equator, across South America, Africa, and the Pacific islands. There are pockets of tropical deciduous forest in India, Sri Lanka, Australia, and the Pacific islands (Figure 53). These forest types, their habitats, and the species that make them up have fluctuated with climatic changes during the geological past.

Because of the close tie between temperature and rainfall, and forest type and growth, we can predict the potential consequences of the expected temperature warmings and changes in soil wetness.

Models doubling carbon dioxide in the atmosphere suggest that the temperatures will rise most at high latitudes, so that the climate supporting the temperate forest is likely to move north into the zones now occupied by boreal forest. While changes in the tropics would be smaller, tropical dry forest climate is likely to invade the area occupied by the subtropical moist forest.

Historical studies show that after past climate changes like glaciation and deglaciation, it takes many centuries for plant and animal species to migrate and adjust their geographic range to the new conditions. The length of the time-lag depends on a number of factors, including the way a species reproduces and migrates as well as the rate of soil development. According to researchers, the tree line in Canada would gradually move about 100 km northward for every Celsius degree of warming.

Climate influences tree growth directly, and it may be increased with global warming in areas where temperature now limits growth; but we cannot predict the effects exactly. Climate warming since the end of the Little Ice Age, about 1850, has resulted in higher productivity in Canadian forests. The greatest absolute growth increase would be in warm, southern, and maritime parts of the northern boreal forest, but benefits could be more than offset by predicted soil dryness. Furthermore, trees growing in warmer temperatures are likely to have higher respiration rates, and that would possibly lower the total net growth.

Trees generally live a long time, up to hundreds of years, so they have evolved with greater tolerance to a wider range and variability of climatic conditions than the short-living annuals. However, we must look at rate of change in relation to past evolutionary experience of trees. Temperature changes anticipated over the next fifty to seventy-five years are much larger and faster than during the past 100,000 years. In the temperate zone particularly, we have yet to assess the hardiness of individual tree species, their ability to survive under different temperatures, soil wetness, extreme episodic events such as early spring thaw, and late frosts, etc. In the meantime, billions of dollars are currently being spent on reforestation programs, where the genetic characteristics of seedlings are carefully matched with the

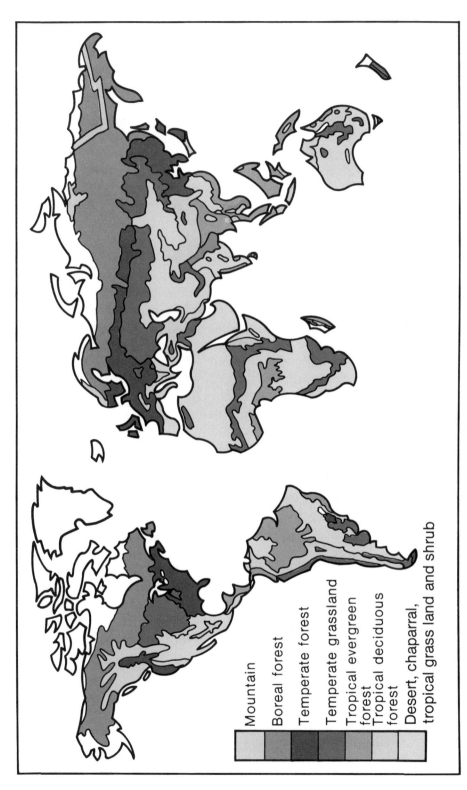

FIGURE 53. *World distribution of forest types.*

current climate of the area being reforested. In Canada, for example, seedlings being planted today, at considerable cost, will be only half their commercially harvestable age by the year 2030, when they could be growing in a very different climate. We need to develop strategies to protect this investment—for example, by developing technologies to harvest and use smaller trees.

Climate change, species migration, and exposure to new habitats create conditions for isolation of species as well as for the evolution of new ones. Climate change would become an evolutionary force that would eventually result in new species, varieties, and forms.

Forest fires, an integral component of many forest ecosystems, would likely increase in intensity and frequency. The activity, numbers, and distribution of many insects and diseases would probably also be changed. The current cold climate of Canada's northern forest zone curbs the northward extension of some insect species, like the gypsy moth, which would otherwise cover the distribution area of deciduous trees, its favoured hosts. Some insects and diseases would surely follow higher temperatures northward.

One of the symptoms of species under stress is reduced or lost reproductive capacity. Species that reproduce by seeds only are likely to be stressed more than those, such as poplars and basswood, that can also reproduce by vegetative means, such as root and stump sprouts. Under global warming, this type of species is likely to survive and to increase in abundance, at least initially, changing the structure and composition of forests. Ground vegetation may benefit from the warming, and compete more strongly with the seedlings of economically desirable species.

Forest-harvesting operations and technologies in northern latitudes would have to adjust to reduced snow-cover and warmer winters, since the present longer forest-harvesting period depends on frozen ground.

The quality, quantity, and seasonal pattern of water-flow in river systems, and consequently the availability of water supply for industrial and residential use, would probably be influenced if watersheds are stripped or the tree-cover is altered or sparse. Altering hydrologic regimes would also change the habitat of forest wildlife.

## HUMAN ACTIVITIES AND SOCIO-ECONOMIC IMPLICATIONS

The relationship between a society and its forests at least partly depends on the society's level of economic development. Commitment to maintaining forests usually evolves with economic development from subsistance to cash economy and then to market and developed economy, as in the industrialized countries of the

Western world. Billions of dollars are being invested in reforestation, management, and harvesting of forests, and in forest-based industries. In Canada, for instance, 350 communities largely depend on forest-based economic activity. The Canadian forest sector employs 270,000 people directly and nearly twice as many indirectly. In 1987 the value of forest products shipped amounted to $36 billion, sixty per cent of it exported—a significant contribution to Canada's balance of payments in international trade. About a fifth of the value of standing stock of $300 billion in the Canadian manufacturing sector is in the forests sector, which invests nearly $6 billion annually in machinery, equipment, and construction. Changes in the Canadian forest resource base would have far-reaching economic consequences.

*Half the world's wood is used as fuel—80 per cent by developing countries, and much of it, as in this cooking fire in Mali, very inefficiently. (Michel Dompierre, CIDA/ACDI.)*

*A Sudanese woman demonstrates a fuel-efficient stove that consumes only half the fuel of traditional cooking fires. Adoption of this technology would slow the rate of deforestation. (CARE photo.)*

*Logging debris on a cutover area in British Columbia, which produces more than 50 per cent of Canadian timber. (SSC/Photo Centre/ASC.)*

Two considerations are important in determining policy responses in any country: first, how we respond to crises; and second, international collaboration.

We can adopt one of at least three strategies in responding to crises: escape, cope, or adapt. In an altered climate, people in a forest-based subsistence economy could likely employ the escape strategy and migrate to other areas; for instance, shifting cultivation in the tropics or becoming environmental refugees. However, this is not feasible for Canada's highly developed economy, with heavy capital investment. Our policy options would be limited to short-term remedial measures (coping), and in the long term adapting to the altered climate regime. Lead-time to make changes and adjustments is most important, and yet that is likely to be short. If the current estimates of the scientific community turn out to be correct, then in less than twenty-five years we will have to deal with a range of new and complex issues around forest insects and diseases, forest fire, vegetation competition, and adaptation of harvest machinery and equipment. Compared to other sectors of the economy, we need a longer lead-time to address issues that determine the growth and the very survival of tree species. This is

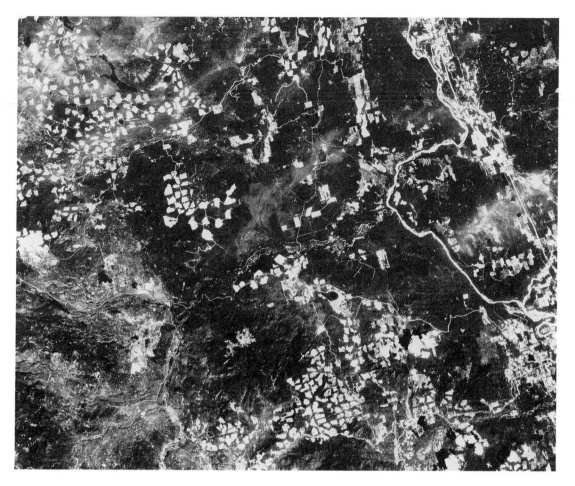

*The white patches show clear-cut forests near Quesnel in the interior of British Columbia. (LANDSAT image courtesy of Canada Centre for Remote Sensing: Surveys, Mapping and Remote Sensing Sector, EMR Canada.)*

a global issue—touching all nations, classes, and peoples—and we need collaboration and collective action by the global community to address it. So far, discussion of this issue has been largely confined to the national and international scientific communites; its emergence on the agenda of the international policy community is relatively recent.

The environmental community has limited experience in formulating international environmental policy and policy instruments, and in establishing international institutional arrangements. Most of our experience so far with multinational collaboration on environmental issues is related to wildlife conservation, the International Joint Committee on boundary waters between the USA and Canada, and more recently to the Montreal Protocol on the Ozone. We can learn from the successes and failures of international environment policy initiatives (for example, the way the Law of the Sea Conference issue was managed from scientific and policy perspectives, as spelled out in Chapter 15), as well as from international collaboration in military and trade alliances. We seem to be entering an era of environmental alliances for global environmental security. Extra-institutional bodies of the 1980s—such as the Palme Commission, the Brandt Commission, the Brundtland Commission, and the InterAction Council (composed of many former heads of state)—have examined complex global issues and formulated alternate visions on, respectively, disarmament, north-south dialogue, environment and development, and forest decline.

It is true that we need collective global action to reduce the emission of greenhouse gases—by, for example, influencing energy policy as well as production technologies and consumption patterns of individual nations (see Chapter 2). Massive tree-planting programs would serve to fix and store carbon for long periods and somewhat dampen the rate of atmospheric change as well. The US Oak Ridge National Laboratory calculates that the creation of 700 million hectares of new forest, a little less than the size of the US, would moderate the

## Time Horizons

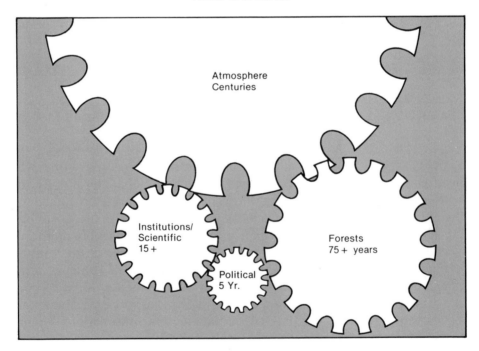

FIGURE 54. *Subsystems involved in forests and atmospheric change operate within different time horizons.*

warming trend of the greenhouse effect. The UN Environment Program estimates that it would be possible to plant 500 million hectares of new forest over the next two to five decades, and that this may be the least costly way to deal with the problem.

We need more research into the likely future response of the forest and human subsystems to altered climate, and how they perform under optimum as well as stress conditions. Examples: responses of human and biological systems to El Niño (see Chapter 2), responses of biological systems to the warm winter Chinook winds in the Rocky Mountain region and to similar phenomena elsewhere, and reinterpretation of the result of growing the same genetic material in different climates.

Policy response is complicated by the fact that the subsystems involved in forests and atmospheric change operate under quite different time horizons (Figure 54). Atmospheric change extends over centuries; investment in forest resources over decades; scientific studies and experiments also take a few decades; institutions like government departments and industry usually plan for ten to fifteen years, though the political time-horizon is usually about five years or less. Forest growth and climate change are long-term phenomena, longer than political decisions allow for. All these gears have to mesh before we can answer the serious challenges before us.

# Brazil

Tropical forests act like a sponge, soaking up rainfall and releasing it slowly and steadily into great rivers and the atmosphere. In doing so they regulate climate—not only local and regional, but global. They cover only about seven per cent of the Earth's surface, in a green girdle undulating about 10 degrees north and south of the Equator. But despite the relatively small dimensions of this band compared to the more northern forests, they contain the most complex and diverse ecosystems, and almost half of all the growing wood, on the planet. Brazil's lush tropical forests depend on diversity to survive. The higher the numbers of species, the greater the genetic diversity, and consequently the more options there are to adapt to changes and adversities.

In the heartlands of Amazonia, tropical forests are being slashed and destroyed daily. Watching developments there is like watching a newsreel of the history of the industrialized

*An untouched rainforest in Brazil. Tropical forests are vital to the planet's climate system, soaking up rainfall and releasing it slowly and steadily. Brazilian rainforests are being razed at an alarming rate. (Pierre St-Jacques, CIDA Photo.)*

*Brazilian ranchers and settlers burn the forest cover for a few years of grazing on the sparse soils. (Pierre St-Jacques, CIDA Photo.)*

world—on fast forward. First we see natives in a subsistence economy, living sustainably within the forest ecosystems. Then Europeans arrive, at first with only minor impact on the forests, later clearing them for agriculture, settlements, generation of capital. And then comes industrialization—powerful technologies transform the landscape at a faster pace and a larger scale. Finally, the realization of what is happening: destruction of the key function of the forests in maintaining the equilibrium of the Amazonian ecosystem. This realization dawns just as we become aware of large-scale global change, and learn about the interaction of all areas—that what happens in the Amazon will affect northern Canada, Europe, Australia.

If the current pattern of destruction continues, by the year 2000 most of the Amazon forests will be completely cleared—over 80,000 square kilometres were burned in 1987. The major motivation is cattle ranching, but the scanty and nutrient-poor soils supporting cleared pastures are depleted in only a few years. Timber extraction and the planting of cash crops like cocoa, rubber, and sugar cane—as well as mining, oil prospecting, and hydro-electric power development—have also been part of the push for new roads through the Amazon, which have brought in over 15 million people in the last two decades. This migration and development has been seen as a solution for internal social and economic problems, as well as a response to the external debt, which motivates the government of Brazil to liquidate its biological capital and to export it to earn foreign exchange (see Chapters 13 and 15).

*In the rainforests trees are often not harvested but left to be burned or to rot. (Pierre St-Jacques. CIDA Photo.)*

CHAPTER 9

# Grasslands into Deserts?

JOHN STEWART and HOLM TIESSEN

People have lived in grasslands for thousands of years—our Western civilization was born 7,000 years ago in the rich grasslands of Mesopotamia, between the Tigris and Eurphrates Rivers in Southwest Asia. Many of the crops we cultivate today came from this area, but the region is extremely fragile. Humankind has learned over the centuries how difficult it is to sustain cultivation on such land, and is finding it even harder today under increasing population pressure and a changing climate.

Centuries of overgrazing and salinization turned the rich Mesopotamian grasslands into desert. This pattern—the rise and fall of a productive farming culture—has repeated itself again and again in our history, and today we see stages of the same cycle in grasslands around the world. In North America, in Africa, and in South America, for instance, people are struggling to devise ways to survive in harsh environments with unreliable rainfall. In the inhospitable boundary regions between grasslands and deserts, nomadic herders have traditionally handled seasonal and cyclical climate change by migrating regularly, while in more moist areas settled farming communities have developed technologies to maintain food production. People tend to settle in such dry environments for the grazing that grasslands usually offer; arable agriculture and animal husbandry combine easily where the vegetation naturally supports grazing animals such as antelope or bison. However, their very coping techniques often lead to deterioration; the results of drought are exacerbated and productivity declines.

Fragile grassland regions are a doubtfully renewable resource already drastically affected by human use. Whatever their future, it will have repercussions on the international food supply and on the life of communities and cultures, and their fate is intimately tied to the uncertainties of global change.

## CHARACTERISTICS OF NATURAL GRASSLANDS

Natural grassland occurs where the environment is too arid for the development of closed forests but not dry enough to prevent smaller plants from forming a dense growth. Precipitation in most temperate grasslands ranges from 250 to 750 millimetres a year, and in tropical and subtropical grassland savannahs from 600 to 1500 millimetres. Drought is a function of both scarce rainfall and high temperature. Rainfall may be higher in the warmer climates, but since evaporation is also high, a closed tree cover cannot form. (Figure 55.)

The driest and therefore the most fragile grasslands illustrate the potential dangers for productive regions when their soils deteriorate or the climate becomes drier or hotter. Patterns of dryland use and the risks they entail repeat themselves around the globe, as shown by the agricultural histories of the North American Great Plains, the African Savannahs, and the Caatinga of Northeastern Brazil. Each region is made up of a series of zones that range from sub-humid to arid—the last with the least diversity, the least resiliency, the least resistance to environmental shocks, both natural and human-induced, and over all providing the greatest problems when cultivated. Moreover, within each region unwise agricultural use has tended eventually to push the zones toward the arid end of the sequence. When marginal grasslands have become desert, a similar trend in adjacent areas converts them from sub-humid to semi-arid. As an area gets drier and hotter, resiliency decreases and therefore fragility increases, as does susceptibility to the shocks that global climate change may bring. The phenomenon is global, with similar patterns and principles underlying the changes in the grasslands and the plight of the people living in them. Although the three regions we are looking at are all made up of these zones, they are less clear in the Brazilian Caatinga, which is itself the driest area of a larger sub-humid region.

Grasses and other grassland plants are tough, and have adapted to withstand adversity. Some survive drought by taking moisture from the soil long after species from wetter regions would have wilted. Others become dormant before desiccating. All produce finely branched root systems, a dense network through the soil, sheltering below ground from the climate above as much as eighty-five per cent of the total biomass. Part of the root system dies every year and decomposes into soil humus, accumulating large stores of organic material underground. The remaining living parts, protected by the soil, send up new growth after drought or fires. This accumulation of organic matter is greater in temperate climates, where decomposition is slower than in the tropics.

Low precipitation determines the nature and extent of natural grass-

FIGURE 55. *Grasslands around the world, showing three areas discussed in this chapter.*

lands in two ways: directly by limiting moisture, and by affecting soil properties. Grassland soils are usually less leached and more fertile than forest soils in similar latitudes, since there is less water to percolate through the soil and remove soluble nutrients. When they are tilled, the accumulated nutrients are gradually released. This is the basis for the grasslands' natural fertility, responsible for some of the most fertile soils known. Long after cultivation has destroyed natural plant cover, these soils continue to produce annual crops.

Although superficially all grasslands look alike, they vary regionally and locally in the vegetation they support. Among the causes are differing climates, topography, soil, and micro-climate. Sustained droughts may shift the boundaries between belts of different vegetation, or change the size of patches in the vegetation mosaic. Wet areas become drier and available for invasion; patches of shortgrass may expand from sandy locations.

Grasslands are generally open, and therefore have higher than average wind speeds. With recurring droughts these can precipitate erosion, particularly on sandy soils. The fine root system binds the soil into a relatively erosion-resistant structure, but once a continuous grass cover is stressed by drought or overgrazing, and gives way to patchy vegetation, then a cycle of deterioration is often inevitable. On the other hand, denser vegetation, or even trees, will colonize previously degraded areas in long wet periods.

Fire, destructive to trees and shrubs, encourages the maintenance or spread of grasslands. Lightning fires periodically burned over huge areas before agricultural communities controlled them, and hunting societies have always burnt over grazing grounds. The aspen forest adjacent to the Canadian prairies, for instance, grows on black soils more typical of grasslands. It probably invaded a region that, up to the last century, was burnt by natives regularly to maintain a grazing range for bison. In tropical savannahs, men often use fire to control pests and weeds or to hunt. Fires are so closely related to the dry grassland climate that we cannot separate them from other climatic factors, and some ecologists describe all grasslands as fire induced.

Almost all the original grasslands of the world have been cultivated, substantially reducing their organic contents. When the sandier soils in drier areas lose organic matter, they become susceptible to erosion; no longer productive, some have to be abandoned, or returned to managed animal pasture. New farm-management practices maintaining a permanent cover on the land through no-till seeding can stabilize the soils, and with prudent management balancing inputs and outputs they attain a lower but stable new level of organic matter within 50 to 100 years of cultivation.

In the Middle East and Mediterranean areas, centuries of over-

*Farmers in many African countries traditionally light bush fires to clear their land, a practice that encourages the spread of grasslands. (World Bank Photo.)*

grazing and inappropriate cultivation practices destroyed the soil-resource base and transformed the original lush lands into deserts. In recently colonized grasslands of Central Asia and Australia, farming is still too new to show long-term trends. In North America the fertile, organically rich topsoil, built up over thousands of years, has been systematically 'mined' for nutrients during the first hundred years of cultivation, depleting up to fifty per cent of the organic matter in the more fragile, sandy, textured soils. The fertile topsoil is a slowly renewable resource, and its depletion threatens the basis for farming and living in the region.

## THE NORTH AMERICAN GREAT PLAINS

The semi-arid Great Plains of North America comprise 250 million hectares. The best land, 105 million hectares, is now cultivated. The rest, mostly marginal for till culture, remains in grassland. Agricultural technology—including fertilizers, biocides, and high-tech machinery—has resulted in tremendous food production, not only satisfying local consumption but also allowing huge exports that have sustained the North American economies and earned for the area the description 'the bread basket of the world'. This label vies with another, 'the great American desert', which was applied during the drought and dust storms of the 1930s. Both descriptions contain some truth, depending on the climatic cycle. Recent experiments with soil-conservation farming methods show that these can stabilize some soils even during short drought periods. The major questions are whether wider fluctuations in precipitation, temperature, and wind, predicted under a changing climate, will lead to further soil degradation and the eventual irreversible destruction of the soil resources of the region; or whether the system will flip to a new sustainable equilibrium, which may not be as productive.

The Great Plains extend through the centre of the North American continent, crossing the boundaries of Canada and the United States and touching the border of Mexico (Figure 56). As in other grassland regions, we can recognize different soil zones and types of plants. In the southernmost belt of the Canadian prairies, for instance, where scarce rainfall and greater evaporation limits productivity, shortgrass prairie steppe occupies barely leached brown soils containing little organic matter. The next zone is dominated by prairies in which short western wheat grass and spear grasses mix with dwarf shrubs such as wild roses and snow berries. Further north, where there is more rainfall and lower temperatures, fescue prairie pastures and aspen groves mark a transition to the forest zone, with deep nutrient-rich black soil containing some of the highest organic matter found in mineral soils.

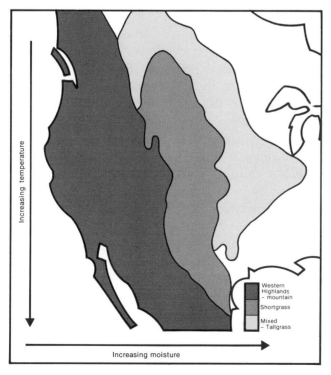

Western Highlands – mountain

Shortgrass

Mixed – Tallgrass

Increasing temperature

Increasing moisture

FIGURE 56. *North American grassland types east of the Rockies.*

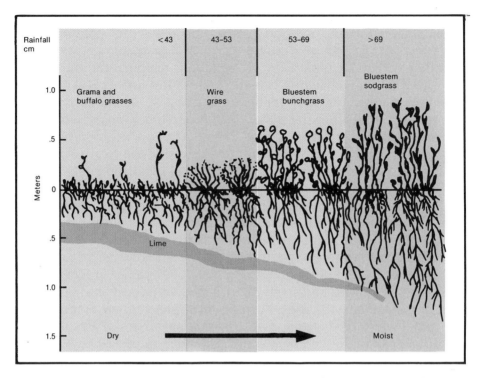

FIGURE 57. *The change in plant communities in relation to changing moisture, from west to east across the Great Plains.*

On the moist eastern margin in southern Manitoba, a tallgrass prairie zone on deep fertile soils is completely cultivated. (Figure 57.)

Early settlers on the Great Plains in the second half of the nineteenth century quickly found they could not transpose the century-old farm-management practices of their more humid European homelands. In the driest areas they grazed cattle, but overstocking caused serious problems until the animals starved during the long winters, reducing herd sizes. The low rainfall would not support economical wheat production. Within a few years the settlers developed a type of land management that emphasized grain-growing, with cattle-ranching as a sideline. In the cultivated areas they introduced summerfallow or cropping only in alternate years to conserve moisture, lessening drought risks and increasing and stabilizing wheat yields. Summerfallowing also released more nitrogen from soil organic matter, further increasing yields. This became one of the dominant management strategies of the Great Plains.

But there has been a cost in moisture and nutrients as a result of summerfallowing. Two-thirds of the nitrogen and some of the phosphorus liberated are not used by the crops, but simply lost. The soil's naturally high fertility is wasted and must be replaced by fertilizer. Erosion because of reduced organic matter and lack of crop-cover has established a vicious cycle: soil fertility is depleted, plant growth declines, organic matter returned to the soil decreases, and so does productivity. In the worst-case scenario—that is, with sandy soils in dry areas, and inappropriate management—the resources built up over thousands of years can be destroyed in less than one hundred.

New management practices were introduced, such as strip farming, and planting shelterbelt trees to lessen wind speeds and prevent erosion. New cultivation equipment to maintain a protective cover of straw or stubble, and new herbicides to permit cropping without tillage, were developed. These technological solutions are expensive, however. Moreover, short-term economics support the strategy of maximizing yields without considering the long-term risks.

It is often hard to tell the difference between the loss of agricultural productivity caused by soil deterioration and natural losses due to changes to a drier climate. Further, advances in technology conceal

*Soil erosion in Saskatchewan after 100 years of cultivation.*

## Climate scenarios with doubled CO$_2$
### (June, July, August)

### Increase in surface temperature

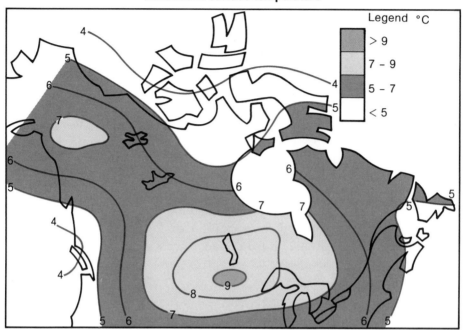

Legend °C

> 9
7 – 9
5 – 7
< 5

### Decrease in soil moisture

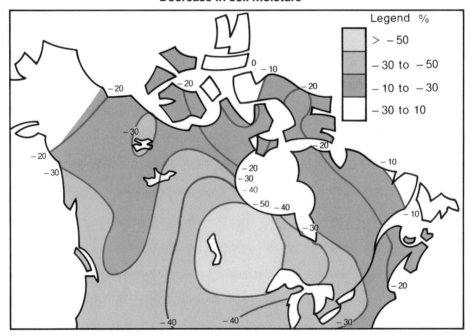

Legend %

> – 50
– 30 to – 50
– 10 to – 30
– 30 to 10

FIGURES 58A *and* 58B. *Predicted surface temperature (a) and soil moisture (b) changes in North America in June, July, and August if carbon dioxide doubles. (Manabe and Wetherald, 1986, US National Oceanic and Atmospheric Administration.)*

the reduction in soil resiliency by keeping production high, or even increasing it. Human-caused and natural degradation reinforce each other, particularly in erosion, and even slightly reduced or more variable rainfall can further endanger semi-arid agriculture. Only the most fragile areas will become unproductive at first, but marginal productivity reductions and erosion increases will be general.

Low moisture has severely limited wheat yields in, for instance, southern Saskatchewan, one of Canada's major wheat-producing areas; yields are only a third of those in more humid France. If they decline further with soil deterioration or a drying and warming trend in climate, wheat farming will become precarious, even with prudent management. Future climate change with greenhouse warming (see Chapter 6) has been modelled in a coupled atmosphere-ocean calculation by GFDL. If carbon dioxide in the atmosphere doubles, hotter temperatures in June, July, and August will result. They could be as much as 9 degrees centigrade hotter in a region straddling the Canada-US border south of Lake Winnipeg. Summer soil moisture would decline from 50 to 30 per cent in all of the North American great plains, with the highest losses in Manitoba and Minnesota. The outlook is not good: a changing climate cannot be counteracted locally, and a deteriorating soil resource will not likely be restored, since the cost would not be returned in a short-term economic system. Economic constraints are even greater in less-developed regions of the world. (Figure 58.)

## AFRICAN SAVANNAHS

At the end of the last ice age the Sahara received enough rain to support vegetation and elephants, antelopes, and other herds that roamed the centre of the continent. This paradise was gradually transformed by climate change into desert, bordered on the south by savannahs. The transformation took a long time, but deforestation, and the loss of plant and animal species from large parts of the arid and semi-arid zones, have drastically accelerated the desert-spreading trend in the last century.

There are three broad vegetation belts between the rain forests and the deserts (Figure 59). The first, the Sahel savannah, south of and bordering the desert, is the driest, with clumps of grass interspersed with small trees and often thorny shurbs. Annual precipitation is less than 350 millimetres. This belt crosses the entire width of Africa, from Senegal to northern Sudan and Ethiopia. Further south, the Sudan savannah belt, with up to 900 millimetres precipitation per year, has a more continuous grass-cover with widely spaced acacias and other trees, like the baobab, typical of semi-arid Africa. The southernmost

*The same tree, the same scene, in the rainy season and the dry season in the West Africa savannah.*

## African Savannahs

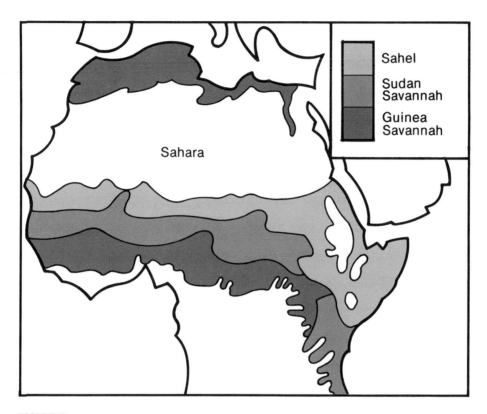

Sahel

Sudan
Savannah

Guinea
Savannah

Sahara

FIGURE 59.

belt before the rain forest is the Guinea savannah. Rainfall here reaches 1300 to 1400 millimetres a year; grasses may reach a height of 2.5 metres, and trees grow in open but almost continuous stands that look like parkland. Natural plant diversity here is much greater than in the drier Sahel; there may be 2,000 different species within a 50-kilometre radius. Repeated burning—part of traditional land management and hunting throughout West Africa—prevents forest encroachment on the southern edge.

A typical feature of tropical semi-arid climates is irregular rainfall. Rainy seasons that start late or end early, dry weeks during the rainy season, or short, excessive rainy periods, may be enough to cause a bad crop year. The semi-arid Guinea savannah has a pronounced dry season of four to five months. During the rainy season intense rains, more than the soil can absorb, cause extensive run-off and ponding. Local farmers respond with elaborate strategies, particularly intercropping (Figure 60). They grow many different crops and preserve some trees, protecting the soil from wind and heavy rains and also contrib-

**Pattern of Intercropping on a Planting Mound in West Africa**

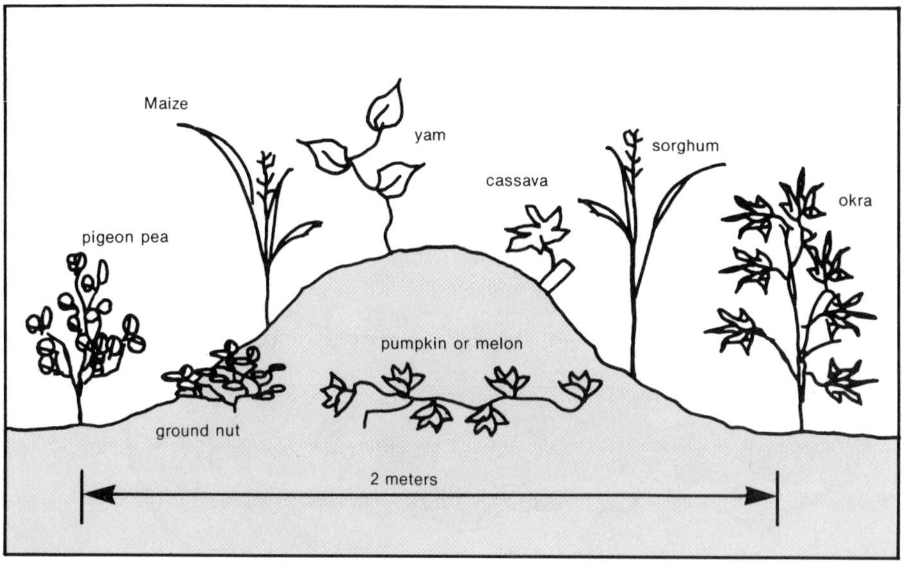

Maize

yam

sorghum

cassava

okra

pigeon pea

pumpkin or melon

ground nut

2 meters

FIGURE 60.

uting to soil fertility and supplementing farm income with wood and food products. It is this diversity that has made their survival possible—and only manual agriculture can cope with the complex, diversified, adaptive cropping systems on which the farmers depend.

Closer to the Sahara, average annual rainfall is lower and rains are even more irregular. In 1985, for instance, two successive plantings of millet dried up in Burkina Faso, and after the third planting seedlings were washed away by heavy rains. In the same year trucks bringing drought-relief supplies into Chad from Nigeria got stuck in the floods caused by the first rains. Here only drought-resistant crops, such as millet and groundnuts, will grow, and with such lack of diversity famine is likely to follow a crop failure.

The main crop diversification is livestock. Cattle, sheep, and goats graze on grasses, tree foliage, and crop residues. The herds are a form of crop insurance; in years of crop failure, the farmer hopes to survive by selling his cattle. Any agricultural or aid agencies trying to limit the herds and prevent overgrazing will first have to devise a new and reliable safety net.

In the driest regions nomadic tribes migrate with the weather, living off animal production and some native plants. In dry years large herds overgraze. They concentrate around the drying wells, transforming the area into wallows sometimes up to tens of kilometres in size. Drilling more wells results in larger herds, which produce disastrous consequences during droughts. Between 1930 and 1970 the livestock population of the Sahel doubled to 24 million, only to collapse during the droughts of the 1970s.

Firewood consumption adds to the problems. In the less-arid Guinea savannah, selective branch pruning keeps trees alive for many years, but the smaller northern trees are normally felled, completely deforesting neighbouring population centres and thinning tree-stands over the entire region. Goats or sheep eat any saplings, making regeneration of tree-stands rare. As a result Ethiopia, a country that once had forty per cent forest-coverage, today has three per cent. This massive tree-loss has lowered water levels permanently in some areas. Improving wood-burning efficiency by introducing simple stoves could probably halve wood consumption.

Traditionally the agricultural land is cultivated in rotation. When the land is broken, native vegetation is slashed and burned, the ashes fertilize the first crops, and the nutrients stored in the soil are released. Since organic matter decomposes quickly in the hot climate, soils are rapidly exhausted, and without fertilization, fields are traditionally abandoned after about five years. Then over twenty to twenty-five years, with a slow succession of weeds, nitrogen-fixing

*Cattle are a form of crop insurance. In years of crop failure they may be a source of income, but overgrazing devastates the land. (World Bank Photo.)*

legumes, and finally a normal healthy savannah vegetation, organic matter and fertility are naturally re-established, and a new cycle begins. This cycling has been successful for hundreds of years, but it means that only a fifth of the land can be cultivated at any one time. Improved health-care has increased the population and therefore the demand for cultivated land, reducing the rotation period. When the answer is to reduce fallow periods, the shortened cycle deteriorates the land.

The result of all these assaults is desertification, a transformation from one ecosystem into another that is more arid, not necessarily because of climatic changes but as a result of soil deterioration. Semideciduous forest is transformed into Guinea savannah, and overgrazed areas in the Sahel become nuclei for wind erosion; dunes form and migrate, spreading the desert. This increases the reflectivity of the earth's surface and reduces rainfall. These effects are an intricate link between degradation and local or regional climate.

Both ecosystem degradation and recovery can be very rapid. Large areas in East Africa shifted from grass to tree savannah in the 1890s when the first outbreak of the rinderpest disease eliminated ninety per cent of the cattle and native ruminant populations. Eradication of rinderpest in the 1960s reversed the process, and tree savannahs in national parks have become impoverished grasslands under the pressure of large elephant populations, which grazed on the trees.

The impact of an increase in human population is even more severe. The population of Africa doubles approximately every twenty-two years. Total food production has gone up, but per-capita food production has dropped twenty per cent over the past decade. In a desperate bid to maintain food production, farmers have reduced fallow periods and practically eliminated them in the highest population-growth areas. The inevitable soil-fertility decline not only reduces food production, but also prevents successful regeneration of natural vegetation when the land is finally abandoned. With no soil-cover, erosion and degradation become irreversible. Abandoning traditional patterns of low-intensity use often reduces rather than increases productivity.

## THE CAATINGA OF NORTHEASTERN BRAZIL

Central Brazil is entirely covered with savannah—the Cerrado, which is semi-arid, but has a relatively even and adequate moisture supply (Figure 61). Agriculture is mostly limited by the acidity and poor nutrients of the ancient soils. On the northeastern edge, rainfall is much less and the tree savannah gives way to grass and the thorn-bush steppe of the Sertão, which receives less than 800 millimetres of annual precipitation, and in some areas less than half that. Rainfall is

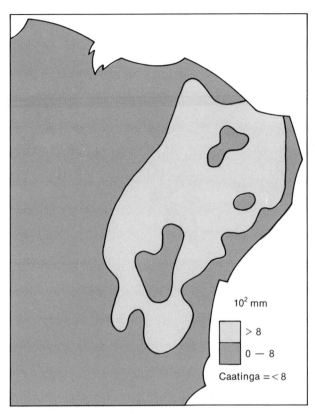

FIGURE 61. *The Brazilian Caatinga, with annual rainfall averaging less than 800 millimetres. The three darker patches are of higher elevation and, like the surrounding area, have more rainfall, producing humid savannah or rainforests.*

$10^2$ mm

> 8

0 — 8

Caatinga = < 8

unreliable and determined by unpredictable global events, such as shifts in the paths of the trade winds, or the El Niño. (See Chapter 2.)

The natural vegetation varies from closed, dense bush to very open mixtures of bushes, cacti, and grasses. The steppe is used for extensive cattle and small-animal production for meat, and subsistence agriculture. Stocking rates are low, 10 to 15 hectares per animal, and cultivation is shifting, except in the irrigable valley bottoms. Traditional crops are beans, maize, and some cassava and cotton. Some fruits, like umbu, are also harvested.

The population is mostly European immigrants, who brought their own traditions, and they and their descendants have practised ranching and relatively undiversified maize and bean production. As in the African savannah, agriculture has to survive on low-input systems, since the benefits from fertilization and other measures always depend on uncertain rains.

Droughts, like those of the early 1980s, cause major hardship and emigration. The Brazilian government plans large irrigation schemes, but by far the greater portion of the Sertão will have to rely on improved management of rain-fed agriculture. The local population has not increased significantly, despite high birth-rates, because of

*The natural vegetation of Central Brazil's savannah includes thorn bushes, cacti, and grasses.*

migration into other parts of Brazil or to the wetter industrialized northeast coast—a cause for social and environmental concern.

In the quest for survival in the Brazilian grasslands called Caatinga, during the past twenty years agriculture has diversified, with small areas concentrating on vegetables such as onions, tomatoes, and carrots; others have been producing specialized goods like latex, wax, or oils from natural or introduced species. Valleys grow upland rice, and the limited irrigated areas produce a large variety of crops.

Despite non-intensive land-use, periodic droughts in cultivated areas have caused local land degradation, aggravated by erosion when the rains return. Desert nuclei, locally known as *deserticos*, have formed, although desert is not a natural ecosystem in the area.

## SOLUTIONS?

The African Savannahs are being stressed by increased demands for food and fibre production, and by recurring droughts that may become more frequent. In Brazil, diversification has not been totally successful. The North American Plains produce huge amounts of food, mostly for export, and are losing their natural fertility as a result. Their productivity is also limited by scarce rainfall, which climate changes seem to be further diminishing. Global climatic models predict even less precipitation in the future. What are the options for improving, or at least maintaining, the productivity of the dry grassland regions, and what are the lessons that can be learned from the three regions we have looked at?

The introduction in many parts of the world of improved varieties, particularly hybrids (the 'green revolution'), multiplied crop yields, so that today many regions are more productive in terms of area-yield than fifty years ago; but this increased productivity has called on ever higher levels of technology and agrochemicals. In many places the great leap in productivity caused by the green revolution has now levelled off, and continued successes are due to slower, painstaking improvements in disease-, pest-, and drought-resistance of crop varieties. To achieve these, plant-breeders often cross tough native semi-arid varieties with high-yielding varieties. Thus preservation of the species diversity found in the dry regions of Africa and particularly in the Middle East—which have been described as the cradle of many of our crops—is vital.

In the African Guinea Savannah traditional product diversification is relatively successful at reducing variable climate risks, but shortened bush fallow periods decrease fertility. Copying the natural ecosystem may be more successful. Integrated agroforestry would maintain a tree population like the natural one and grow diverse herbaceous crops in place of the native grasses and shrubs.

Diversification has to some extent taken place in the Brazilian Sertão, where European traditions have led not only to a damaging subsistence monoculture, but also to more flexible land-use patterns. Farmers grow a large variety of products: wood fibre, fruit, secondary metabolites such as waxes, resins, latex, or pharmaceutical raw materials, and adapted field crops or vegetables. Frequently, however, other factors override success in the marketplace. The naturally low semi-arid productivity makes competition with more productive areas uneconomic, and international oil prices control the market value of latex, fibre, and oil crops, etc.

Similarly, in the Great Plains of North America, economic and marketing goals perpetuate a production system that has caused widespread resource degradation, and may totally collapse if current trends are compounded by a drier climate. Diversified markets, and production tailored to them, may be a partial solution. It would be possible, for instance, to sell range animals to Europeans and others who are concerned about the health implications of excessive chemical, antibiotic, and hormone use in their own industrialized meat production. Again, it might be possible to raise animals, such as bison or wild game, that are adapted to the semi-arid plains, and to grow a variety of plants other than wheat.

Natural ecosystems in the semi-arids cannot be preserved as long as populations increase, or as long as cheap exports of mass commodities, such as wheat, dominate in the less-populated regions. Adapting production patterns to the carrying capacity of the land as well as to economic and nutritional needs is complex, calling for ecologically sound and marketable plant communities. Integrated production systems stressing risk avoidance rather than high productivity—like the agroforestry suggested in Africa—need to be encouraged, but they will probably not support short-term wealth accumulation of more exploitative land uses. Agricultural policies can succeed only if accompanied by complementary social and economic policy decisions.

Much of the land currently in production in the North American Great Plains is not stable; it is subject to erosion and is especially vulnerable in periods of prolonged drought. Three-to-four years of lower-than-average precipitation make the complete region extremely vulnerable. US and Canadian governments have therefore made plans to remove their most endangered areas from production, but an integrated land-use strategy is missing.

Present semi-arid agricultural practices are not as successful as yield statistics imply, even in the most developed countries, because they are destroying the soil. This process is generally still reversible. Fundamental changes in the cropping systems may be initiated by individual farmers, but they must be supported by societies prepared to invest in the long-term land resource and in research based on the concepts of biological and non-biological controls.

Soil deterioration, in the worst scenario, takes fifty to a hundred years. Unfortunately it takes a similar time to see changing management bring improvements. These time-scales are hard for decision-makers to take into account, concerned as they usually are with politics and economics of less than a decade. Nevertheless, we desperately need far-reaching political decisions on land-use based on current knowledge of the processes affecting this fragile resource.

# World Conservation Strategy

*World Conservation Strategy* (1980), which brought about a turning-point in thinking about economic development and protection of the biosphere, was commissioned by the United Nations Environment Program (UNEP) and prepared by the International Union for the Conservation of Nature and Natural Resources (IUCN), with assistance from the World Wildlife Fund and UNEP itself. It sets out three fundamental objectives for the conservation of living resources:

—maintenance of essential ecological processes and life-support systems;

—preservation of biological diversity;

—sustainable use of species and ecosystems.

The plan was developed, and is being promoted and applied, in a very decentralized fashion. Although some countries, particularly in the developing world, have endorsed it, individuals elsewhere around the world in government and non-government agencies are using it as a blueprint and checklist for policy development.

The *Strategy*'s priority requirements make conservation a primary policy concern of all economic sectors in government rather than of one separate department. It designs environmental policies that can 'anticipate and prevent' rather than 'react and cure'. National actions would conserve renewable resources like fisheries, forests, and cropland; protect natural areas; and control pollution. International actions would protect the global commons: the atmosphere, the oceans, and the continent of Antarctica.

Although the project broke new ground, it had a significant shortcoming: it lacked any provision either for promoting the *Strategy* or for monitoring progress. Implementation requires that every political jurisdiction, where actions and policies affect living resources, must develop its own conservation strategy (or, as many now call it, sustainable development strategy). Nearly ten years after the publication of the original *Strategy*, there are fifty-four other strategies under way or completed in forty-one countries, and the IUCN is setting up a data base to keep track of them.

At the national level, a number of countries in Africa, Latin America, and Southeast Asia—where environmental problems are closely linked to poverty and a degrading resource base—

are using the *Strategy* for policy-making. In Canada, where environmental concerns and sustainable development have a high level of public support, almost all the ten provinces and the Territories have committed themselves to producing strategies by the early 1990s. But the larger industrial nations—such as Britain, the USA, and the USSR, with more entrenched economic interests and political structures—find the plan less congenial.

The concept of conservation as 'a primary policy goal of all economic sectors' is a central message of *Our Common Future*, the 1987 report of the Brundtland Commission. Applied nationally in the area of energy policy, for example, this concept would require policy-makers to consider from the beginning the ecological implications of different options, instead of only narrow economic criteria. No country has yet achieved this integrated approach; but with luck, the effect on the environment of whole policies and programs, rather than just individual projects, will in future be assessed before development.

A 1986 international gathering, sponsored by the IUCN, reviewed the content and application of the plan. Participants expanded the goals to address social justice ('conservation with equity'), military preparation and conflict, human population levels, and non-renewable resources. While the *Strategy*'s biophysical requirements for sustainable development are widely accepted, its social, political, and economic dimensions are only now coming into the limelight.

SUSAN HOLTZ

# PART IV

# Managing Ourselves as Part of One World

Humankind today faces some incredible and difficult decisions that few of our ancestors ever had to face. The growth of population, coupled to the growth of knowledge and our technologies, have led us to question if our past strategies can be sustained.

The rate at which we are changing the biosphere and its support systems boggles the mind. Soil is eroding and species are being destroyed. We live in a world increasingly divided between the rich, those who have never lived better, and billions who live on the knife-edge of grinding poverty. Our world is becoming increasingly socially unstable, even intolerable. With modern communication and observation systems, we cannot hide from reality. What will our world look like in 2100?

We must consider the thoughts written by Aldous Huxley in 1948:

Industrialism is the systematic exploitation of wasting assets. In all too many cases, the thing we call progress is merely an acceleration in the rate of that exploitation. Such prosperity as we have known up to the present is the consequence of rapidly spending the planet's irreplaceable capital.

Sooner or later mankind will be forced by the pressure of circumstances to take concerted action against its own destructive and suicidal tendencies. The longer such action is postponed, the worse it will be for all concerned. . . . Overpopulation and erosion constitute a Martian invasion of the planet. . . . Treat Nature aggressively, with greed and violence and incomprehension: wounded Nature will turn and destroy you . . . if presumptuously imagining that we can 'conquer' Nature, we continue to live on our planet like a swarm of destructive parasites—we condemn ourselves and our children to misery and deepening squalor and the despair that finds expression in the frenzies of collective violence. ('Managing', *The Double Crisis*.)

Consider Huxley's significant words: greed, violence, incomprehension. As the Bruntland report so correctly emphasizes, development alone must not be our objective; it must be *sustainable* development. We must learn not to consider simply the short-term advantage for a few. At present, by our violence, we are destroying living species at a rate near the per-cent per-year level. Soon we may modify the genetics of a vast array of species, including our own. Do we have the knowledge and comprehension to understand the long-term impact of our actions? Clearly we do not. The great new scientific endeavour launched by over fifty nations in 1986—the so-called International Geosphere-Biosphere Program: A Study of Global Change—is aimed at understanding the complex chemical, physical, and biological *interactions* that control the life-support systems of our planet. Before we have much better information on this ultimate problem, we must proceed with great caution, recognizing our ignorance of our life-support systems. For almost all human history our focus has been on the anthropocentric management of Earth. Is this a sustainable form of management? Or should we consider what some call an *ecocentric* attitude for management?

WILLIAM S. FYFE

# CHAPTER 10

# Grounds for Concern: Environmental Ethics in the Face of Global Change

## PETER TIMMERMAN

It has now become almost a cliché that everything has changed except our ways of thinking, and that global change is going to require a fundamental reappraisal of how humankind relates to Nature. Yet nothing much changes: our growth patterns continue and our priorities shift at a speed that rivals continental drift. Meanwhile, if we are to believe the scientists, time is the resource we are most rapidly running out of.

How are we to change? Are we to find new values, return to those we have somehow lost, or simply live up to those we now only profess? Can modern society be reconciled to Nature, or are the ways of each in basic conflict? Are our traditional sources of moral and ethical thought so corrupted and outdated that they can be of little use, or can they be refreshed and restored by the challenge of global change? Is radical rethinking necessary, or can we make do by simply being a little smarter and a little more efficient?

These questions—which rapidly move out of the realm of ethical and spiritual questions and turn into economic, social, and political questions—are now being addressed in a growing, but still amorphous, area of study known as 'environmental ethics'. Because of the magnitude of the issues involved, and for reasons that should become clear as this chapter continues, 'environmental ethics' is seen by many of those involved in its study as not just another topic of standard ethics, such as 'business ethics' or 'legal ethics', but as an attack on many of the main assumptions of much of the history of Western thought and experience.

This chapter is an attempt to approach the issue of changing our values by looking at the historical development, up to the present, of 'environmental ethics', and using that as a springboard to go on and discuss some of the current debates and potential implications of a new ethics for the environment, both at the local and global scale. The sketch obviously will be selective, for reasons of space and personal bias towards what I believe is most interesting and challenging, since a full discussion of humanity's relations with nature could encompass the entire historical (and prehistorical) record!

## THE FLOOD AND AFTER

Some years ago, in a book called *After Virtue*, the social philosopher Alaisdair MacIntyre tried to describe our current moral confusion by means of a compelling little story. He asked his readers to imagine that there had been a society in which there were a number of complete moral systems available to which people gave their allegiance. These moral systems gave shape to people's lives, helped explain the presence (or absence) of good and evil in the world, and often had a series of fundamental rules that people who belonged to those systems used to make decisions about what they ought to do. MacIntyre then asked his readers to imagine that this society had been subjected to a flood, a 'moral deluge', in which all that was left of these moral systems were a few islands poking above a sea of chaos. Because these islands were no longer connected to each other or to an agreed-upon mainland, any debates between them would be rootless and often interminable, since the participants would be talking from different fundamental premises, which even those who believed in them would be unable to locate or articulate.

The obvious point MacIntyre wanted to make is that our society is that imagined society; and can be seen as one in which fragments and memories of strong moral and ethical systems survive, but have for the most part been drowned in a sea of relativism or irrelevance. Whatever one ultimately thinks of his diagnosis, it is fair to acknowledge that we live in a society of competing moral and ethical demands. Some of them are based on very ancient, traditional religious or philosophical teachings, and others are quite modern, deriving from such notions as utilitarianism, efficiency, or other analyses. For many people, the only time they suddenly realize that they live in overlapping or competing moral systems is when they face some crisis where one set of values says they should do x, another set of values says they should do y, and another says that x and y are irrelevant to what really matters, which is z. Many of these systems have implicit or explicit ideals about the value of the natural world and how human beings define themselves in that world. And we find when we try and sort out our relationship to nature, that they are usually a confusing hodgepodge of ideas, images, and—as MacIntyre might put it—remnant fragments of ancient systems of belief. One of the most interesting aspects of the rise of 'environmental ethics' is the way in which environmentalism has tried to return to 'the world before the deluge' to seek out some of these ancient systems of belief for renewed moral guidance.

When we look at traditional ways of relating to the natural world, beginning with the aboriginal or native traditions, we find that the spiritual and the practical are strongly interrelated. There is some evidence that certain prehistoric tribes may have operated on occasion in ways that were wilfully damaging to the environment. But the vast majority of ethnographic experience indicates that hunter-gatherer societies and others were exquisitely balanced with their environment, and this balance was due in many instances to extensive detailed knowledge of the natural surroundings and a culturally expressed system of checks and balances, of territories and hunting rules, of times to work and times to play.

The fact that human beings were not in charge of the uncertainties of the world around them, and set out to do something about this, may be one of the most important factors in the rise of human culture. When, for example, we look at the history of religions, we find everywhere attempts to figure out what the gods of nature want, to propitiate them, to wheedle them out of doing bad things and encourage them to do good things. One of the triumphs recorded in the Old Testament, and later in the Koran, is the breaking-out of this enthralment to the gods of nature (the pagan idols of the Scriptures) towards a single universal God 'beyond' nature. It has been argued—and, as I will shortly note, counter-argued—that this has made the Judaeo-Christian tradition inherently biased against uncultivated Nature.

In other religious traditions, most notably in Hinduism—but with variations elsewhere in, for example, certain native American traditions—religious ritual evolved through trying to match the patterns and rhythms of nature and nature's gods. In a broad generalization, the essential task of human beings was to find out what those patterns were, and eliminate those flaws and impurities in their lives that made it hard to 'tune into' or resonate with Nature's harmonies. These traditions thus speak about the 'Way', the 'Dharma', or the 'Brahman'. We find elements of this attitude in certain unorthodox Western traditions, such as Gnosticism, alchemy, and in historic and literary references to 'natural law'. Our literary tradition is saturated with such a world view: many of Shakespeare's works, for example, depend for their poetic resonance on the clash between an older or folk view of how nature trembles when evil is done on earth, and a new world of power politics that ignores such silliness.

Many people date the rise of 'environmental ethics' as an academic study from a famous 1967 article in the journal *Science* by the American historian of medieval science, Lynn White, Jr. In this piece White accused Christianity (or, to be more precise, some aspects of it) of fostering the technological attitude towards nature as an object for exploitation and manipulation. This debate, which raged throughout the 1970s and is now mostly dead, brought into the light both a number of hidden assumptions in medieval and modern history about the images of the human and the natural. It also brought forward a number of interesting counter-examples and counter-arguments. These ranged from descriptions of the environmental destruction wrought by many non-Christian and supposedly benign nature-loving nations, to citations from those parts of Scripture that showed how much the Psalms (for instance) are full of the celebration of God's handiwork in creation.

The inconclusive aspect of the debate stemmed from the fact that the complexities of deciding what caused what in ordinary history is made even more complex when trying to add the influence of ideas and general cultural tendencies into the mix. That the debate has been inconclusive (though most people would agree that Lynn White was on to something of importance) implies, of course, that there is no easy solution, such as: 'Get rid of all remnants of Christian thought, then technology will disappear, and all our ecological problems will go away.'

The best part of this debate, however, was that it stimulated a new generation of Christian theologians, and interested adversaries into making the issues clear and the arguments better, so that some of the best writing about environmental ethics is now coming from Christian theologians and philosophers. Environmental ethics has also stimulated those in other religious traditions to clarify their own ideas about our terrestrial home.

## THE ROOTS OF ENVIRONMENTALISM

Among the facts that are not in dispute is that with the rise of science and its offspring technologies, beginning the seventeenth century, a new way of looking at nature was born.

It has been argued, by philosophers such as Charles Taylor of McGill University, that the rise of science itself brought with it a set of values that had been adopted, or were inherent in the ways in which science went about its work. It needs to be reiterated that, until the rise of modern science, the way in which human beings discovered the patterns of the universe was to 'get in tune' with them, to discover the ways in which the microcosm (human life) and the macrocosm (the universe) matched in meaning. A wise man was often he who was most in touch with the things that mattered. The phrases 'get in tune' and 'most in touch' are remnants of the sense of resonance with the patterns of a meaningful universe. With Descartes and Galileo, the essential trick was to break away from the seductive superficial meanings of the patterned world, so as to clear the mind. Part of the power of science is its use of disciplined experimentalism as a method of reining in our very human tendency to find what we hope to find in the world. This breaking away into objectivity, into separation, has allowed us to make many of the discoveries and human improvements of the past 300 years, most of which no sane person would want to eliminate. There has, however, been a cost: what the world has gained in factual meaning, it has lost in metaphorical resonance.

In addition, of course, or as part of this process, the past 300 years have seen the rise of industrial technologies that have allowed us to buffer ourselves against many of the vagaries of nature, to the point where obvious connections with nature and nature's processes are absent in the artificial cities within which more and more of the world's population lives. When we look for the roots of modern environmental ethics, we find them in the original protest movements against the dark side of the Industrial Revolution. They were not just about the despoiling of nature, but also about the loss of rural community, the growing alienation of human beings from their work, and the struggle for individual dignity in a world of ever-larger physical and bureaucratic machinery. These protests have become woven inextricably into environmentalism, and one can often find in the literature on environmental ethics calls for the protection of the natural world that are disguised as calls for the protection of aspects of humanity. Nature thus once again becomes a 'carrier' of other concerns.

This brings us to one of the central paradoxes of contemporary environmentalism; that is, in the attempts to come up with both an environmental ethic and a new environmental politics. As we have just noted, the roots of environmentalism can be found in the reaction to the rise of industrialism, a reaction that was usually labelled part of the goal of the Romantic Movement to reassert imagination and sentiment and emphasize individualism in thought and expression, rather than the restraints of conventional formality. Romanticism was (and is) a very mixed bag of responses. Some kinds of Romanticism—those concerned with the loss of community, with the loss of tradition— were very conservative and traditional. Other kinds—those concerned with tackling industrialism head on—were more radical. Because of this, there has been a constant to-ing and fro-ing between right and left in environmentalism. The turn-of-the-century 'conservationism' in North America that launched the national parks system was a mixture of both. In more recent history, much of the philosophical and political force of environmentalism has been associated with the radical Left, drawing on those aspects of environmentalism that derive from Marxist thought, anarchism, and the general Left tradition. Yet it can be argued with equal force that environmentalism ought to be seen as deeply conservative, with other roots in such concepts as those of Edmund Burke—who emphasized the organicism of society and history—and with what has been called the 'Radical Tory' tradition of Swift, Ruskin, and others. This tradition is a counterpoint to the Leftist tradition, and can be found in the thought of such diverse figures as Mahatma Gandhi in India, Simone Weil in France, and the British economist E.F. Schumacher and scientist James Lovelock, the promoter of the Gaia hypothesis.

Environmentalism is thus perhaps recapitulating a familiar historical trajectory embedded in the original meaning of the term 'revolution', a return to an earlier cycle, or in the word 'radical' itself—back to the roots. Environmental ethics shows the same tensions, and environmental politics is split vertically between New Left environmentalism and Old Right environmentalism, both of which agree only on the view that the modern industrial state is a pathological menace. One can therefore find throughout the environmental movement odd coalitions of urban anarchists, concerned farmers, and the muddled in the middle. There is also emerging a darker side to this Right/Left split: in the United States there can now be found extremist groups of the Right (gun-toting neo-Nazis protecting their land) and the Left (saboteurs of electric power lines and logging machinery).

## FROM THE SHALLOWS TO THE DEPTHS

At the same time as this Left/Right split, there is also something of a horizontal split between what is called 'shallow ecology' and 'deep ecology'. This dichotomy—first put forward in the works of the Norwegian philosopher Arne Naess in the 1970s—is designed to separate those who believe that we can continue on our current path with only certain (albeit major) revisions to our life-styles and ways of life, and those who believe that only a deep change in human relationships with nature will save us.

With 'shallow' and 'deep' ecology we return to one of the questions with which we began this chapter: is environmental ethics merely an extension or variation of standard ethics, or does it require a completely new approach? This is not simply an interesting philosophical issue: when we look at the now-fashionable phrase 'sustainable development', we might divide the world into those who see it as 'sustainable *development*' (e.g. most politicians, economists, and businessmen) and those who see it as '*sustainable* development' (e.g. many environmentalists, some politicians, a few businessmen). Deep ecologists would repudiate both interpretations and argue that the question 'sustainable development *for whom* or *for what*?' is never asked, because it is naturally assumed that we are talking about human beings, always.

One now-familiar battleground that may be taken as symbolic of many other struggles is the question of 'animal rights'. Most average people have very complicated emotions and images of the animal world, including not only occasional twinges of guilt about meat-eating, but also familial love for pets and curiosity about exotic creatures seen in zoos or on nature documentaries. The limits of their usual response is 'humaneness', as in the Humane Society. When we move into the philosophy—and action—of 'animal liberation' or 'animal rights', much of the argumentation involves an extension of standard human ethics into the natural world: that is, arguments are put forward that animals should be considered as having 'rights' just like people, or that because animals have certain levels of consciousness or can feel pain and suffering, there is no decent reason why they shouldn't be given the same concern as human beings in equivalent cases.

Many deep ecologists, while sympathetic to this approach, would say that having to find those aspects of animals that most resemble human beings, in order to interest human beings in their welfare, shows how perverted our relationship to nature has become. They would instead propose that we need to begin with a re-grounding of our fundamental natural connections to the ecological communities of which we remain an integral part, and that it is only when we have achieved this reorientation that we can know how best to act. As a result, when we come to the issue of whether or not traditional hunting should be allowed, one can occasionally see deep ecologists and animal liberationists on different sides. If there is a theme that runs through deep ecology, it is one of community—ecological communities that see human beings as only one element of a community of relationships that may well extend into the non-living world. Deep ecologists would perhaps characterize themselves as 'ecocentric', as distinguished from 'biocentric', opposed to 'anthropocentric'.

There is no space here to debate the virtues of shallow versus deep ecology (or to do more than skim the surface of the furiously contentious animal-rights debate); only to say that global necessity is becoming the mother of ethical invention. It is becoming clear that we have emerged from a long period where we did not need to take the natural world into account except as a 'given'. We now find that if we keep treating it as a given, it will be taken away from us. In many parts of the world, political theories and governing structures that were created in the nineteenth century, during a period of unrestricted growth and increasing local insulation from nature, are proving to be inadequate in the face of the resurgence of what can be called 'Nature's veto'. It may be that in order to deal with this veto, we are not only going to have to negotiate a 'global contract' among humans (see the Statement following this chapter on the human dimensions of global change); but also, in the words of the contemporary American naturalist Barry Lopez, renegotiate our contract with Nature. In previous societies this contract was expressed in social taboos, conversations with the animal totems of the tribe, or through the long hammering out of a live-and-let-live policy with an often bountiful, but occasionally recalcitrant, sacred Nature. Environmental ethics, in its various forms, may well be seen as laying the groundwork for a revitalised, reciprocal contracting.

Talk of contracts is perhaps too mechanistic or legalistic for the sorts of regenerated relationships now being discussed in the environmental-ethics literature. Some thinkers now argue that any new relationship can be developed only in the context of a new mythology, or new 'cosmic story', which would marry the insights of contemporary science to the rapidly emerging global awareness. If such storytelling is possible—a 're-enchantment of the world', as some are calling it—it will require an unprecedented collaboration between artists, poets, writers, theologians, and scientists. It will also require a recognition that our current language has become so impoverished that in spite of 24-hour news broadcasts, we are unable even to tell ourselves what has gone wrong—that part of our problem is that we have been reduced to gestures of inarticulate anxiety.

Behind this notion of a new global story-telling or re-mythologizing is the understanding that, as human beings, our actions are inevitably based on (and often controlled by) the stories or narratives we feel compelled by, or owe allegiance to. Some of these stories have become tired and outworn. Examples are stories of the endless frontier, the nation over all, growth for the sake of growth, the rugged invulnerable male. But they are hard to dislodge because they have served to give shapes and meanings to so many lives for so long. If we were to find or create such a story, many of the problems we now have to wrestle with explicitly would resolve themselves simply by being embedded in a new 'geo-' or 'eco-philosophical' context. Decisions about where to live, how to make a living, whether to drive to work, and all of our relationships, would be thereby affected.

The potential power of such revisionary stories should not be underestimated. For example, the most powerful movement in the Christian church at present is 'liberation theology', based on what is called 'the preferential option for the poor'. This movement began with a series of re-readings of the Gospels. The first re-reading produced the argument that the poor should be given first consideration when it comes to determining the impact of any given decision or proposed project, but this very rapidly evolved in subsequent re-readings into the far more formidable challenge to see the world and its institutions from the perspective of the poor and the vulnerable—from the bottom up, and not, as usual, from the top down. If there is to be a new global story, it will probably have to take into account the fact that the natural world within which we live now belongs as well to the poor and the vulnerable, and that our story will need to include 'the preferential option for the earth'. It will also have to remember that not all the stories of the earth are told with human voices.

## TO HIGHER GROUND

A Tibetan Buddhist was once asked how he felt about the teacher-disciple relationship, and he replied that the more independent he became of his teacher, the more he realized how dependent he was on this teacher's wisdom and compassion. Indeed, the more truly free and independent we become, the more aware we become of how much we depend on others and how appropriate it is that we appreciate that dependence. For much of the past few centuries we may have become prisoners to a false idea of liberation, one that was profoundly influenced by the need to break free of authoritarian fetters of all kinds, imposed on us by antiquated social structures, outmoded world views, and some of the constraints deriving from ignorance of natural processes. If there is to be a fundamental change in our values, it may be that we are going to have to re-evaluate our most cherished beliefs—in the absolute value of every human life, in absolute freedom, in the good life. The challenge is to undertake such a re-evaluation without throwing away those values that generations of human beings have fought and died for, and without giving up the struggle for the extension of these values to those who have not yet had their full share of them; and yet at the same time admit the possibility that there are problems with them. One example of this dilemma is in the notion of 'rights' as the defining characteristic of human value, while much environmental thought is really not about individual rights, but about duties, obligations, and common responsibilities. Many debates can be rearranged so that it appears that we are talking about 'environmental rights'—rights to clean air, water, and wilderness—but the manipulation is strange, especially when we are driven to give trees and endangered species pseudo-human rights before we will take them seriously.

In the end we may have to come to a position not unlike that of the Tibetan Buddhist: to learn that true independence is a form of dependence or interdependence, that without the microbes and the earthworms, all our posturing would never happen. We will need to learn again a little humility, derived, like 'humus', from a word in Latin meaning 'ground'—from which we also got, long ago, the word 'human'.

# Changing the Globe, Changing Our Minds: Towards a Global Contract

As we enter an era of global change we need a new relationship between human society and the environment that will not just be economically and ecologically sustainable, but morally sustainable as well.

Our growing power over the planet, and the recognition that we live in a new context, exemplified by the pictures of our fragile globe hanging in black space, have altered the political and social chemistry of our time. Around us we see fragmentary attempts to forge rules that will be binding on all members of the global community—based on the principle that we must all hang together, or we will all hang separately.

Perhaps the time has come to set down explicitly, through negotiation, a formal global social contract for the emerging global community.

A global contract in this sense is not only a political contract, concerned with the relationship between governments and people, but it is first and foremost a contract of society. Historians and philosophers are familiar with this concept, since it was a central issue for such philosophers as John Locke, David Hume, and Jean-Jacques Rousseau. To them the concept of a social contract was as much as anything a theoretical method of appraising the current state of a society, but we have perhaps come to a point where a real contract needs to be hammered out. For we may be on the verge of breaking—through nuclear war or ecological catastrophe—our earlier indissoluble and organic contract with all generations of life on earth, past, present, and future.

The basis for such a contract is already in the making, in commonly agreed-upon and generally understood norms of behaviour in much of global society, and in continuing efforts to make nation-states live up to their own codes of conduct. It is embodied in the entreaty to 'think globally and act locally', and in the growing literature on a global-environmental ethic. It is also embodied in a few halting, hard-won international legal agreements, such as the Montreal Protocol on Atmospheric Ozone Depletion.

Treaties and entreaties, however, will not be enough. We need to reconsider the relationship between economics and ecology, between excessive military spending and environmental degradation, and whether or not our current ways of life are, in fact, sustainable for the foreseeable—and unforeseeable—future.

One seldom-acknowledged impetus towards global understanding is the gradual recognition of interdependence, fostered by global scientific research into ecological cycles, climatology, and an improved resolution of planetary cartography. Current global-research initiatives, including the International Geosphere-Biosphere Program and the Human Dimensions of Global Change Program, are part of the groundwork for a global contract. Scientific global co-operation has for many years been well in advance of co-operation in many other spheres, and has contributed to the fact that concern for the global environment seems to be one of the topics where nations and concerned individuals can meet.

On this common ground of concern a global contract may be founded. As in all contracts, it will be doubly binding: it will not just articulate our rights to basic needs and an equitable distribution of the benefits of living on Earth, but it will also spell out our duties and obligations to that Earth, to each other and to the generations yet to come, for which we now bear a heavy responsibility. Like Atlas, we now carry the globe upon our shoulders. Heaven help us if we just shrug it off.

## AN INTERNATIONAL PROGRAM ON
## HUMAN INTERACTIONS WITH THE EARTH

The Human Dimensions of Global Change Program (HDGCP) is an international means of bringing together social scientists, natural scientists, and those involved in the management of human activities in research to be focused on key interactions with the Earth. The steering committee is made up of the International Federation of Institutes for Advanced Study, the International Social Science Council, UNESCO, and the United Nations University.

The rapid alterations to the physical dimensions of the biosphere, described in this book, indicate that the effects of human beings on earth systems can no longer be ignored. These alterations are now beginning to have major impacts on the social, economic, and political structures of humankind. A number of concerned individuals and institutions around the world felt that the time had come to explore the human dimensions of global change.

The impetus for the HDGCP comes in part from the current physical-sciences initiative, entitled the International Geosphere-Biosphere Program (IGBP), operating through the International Council of Scientific Unions. The HDGCP plans a decade-long research effort to explore the dynamics of human interactions with global environmental changes, as well as to provide a basis for improving policy solutions to problems that arise from these interactions.

The HDGCP will operate internationally through a series of Working Groups, monitored by a Scientific Committee, and under the overall direction of a steering committee. At the national level, groupings may be organized through special committees, or in conjunction with IGBP activities. In Canada the national Program will be administered through the Royal Society of Canada, which has a unique capacity, due to its social and natural science membership, to act as a bridge between the IGBP and the HDGCP.

## OBJECTIVES

At the September 1988 Tokyo International Symposium, the following objectives of the research initiatives for the Human Dimensions Program were adopted:

• to improve scientific understanding and increase awareness of the complex dynamics governing human interactions with the total Earth system;

• to strengthen efforts to study, explore, and anticipate social change affecting the global environment;

• to identify broad social strategies to prevent or mitigate undesirable impacts of global change, or to adapt to changes that are already unavoidable;

• to analyse policy options for dealing with global environmental change and promoting the goal of sustainable development.

In pursuit of these goals, the research program will undertake several activities, including initiatives

(i) to foster a global network of scientists and other concerned parties, and to encourage this network—in collaboration with other relevant research initiatives—to engage in research directed towards the dynamics of human interactions with the global ecosystem;

(ii) to undertake selected core research projects central to the purposes of the program;

(iii) to develop appropriate information systems and methodologies to execute a research program of this scope;

(iv) to explore the ethical, cultural, and legal traditions and frameworks that underlie and shape the human aspects of global change;

(v) to propose procedures and techniques for helping the translation of research findings into policy-relevant terms;

(vi) to promote educational efforts about human activities having significant effects on the global environment.

## DEVELOPMENT AND CURRENT ACTIVITIES

The first meeting for the Program was held in Toronto in June 1987. Since then consultative meetings, workshops, and conferences in various parts of the world have explored the opportunities provided by this initiative, and a growing network of interested and committed scholars and institutions is being developed. Out of this active consultative process, six themes for initial pilot projects have been identified. These themes have intrinsic global significance, draw on already existing expertise and institutional support, and define broadly integrative areas within which interdisciplinary research can be undertaken. The working groups are:

1. *Global Climate Change: Strategies of Response*: to, among other tasks, examine the strategic requirements for meeting the targets set at 'The Changing Atmosphere' Conference held in Toronto in June 1988 (see Chapter 2).

2. *Needs of the Most Vulnerable People and Places*: to explore ways to reduce the vulnerability and enhance the resilience of the peoples and places of the world that are most likely to suffer from environmental change.

3. *Global Risk Assessment*: to apply risk-assessment concepts and techniques to global issues, and to expand the role of risk assessment in developing countries.

4. *Industrial Restructuring and the Analysis of Industrial 'Metabolism'*: to assess current attempts to restructure industries in many developed countries, and the stress of rapid industrial development elsewhere.

5. *The Legal, Ethical, and Institutional Dimensions of Global Change*: to explore legal and institutional barriers and instruments for effecting global environmental sustainability, as well as the ethical framework within which global change is being evaluated.

6. *Methods of Analysis, Modelling, and Data Requirements*: to provide a methodological basis for many of the research activities of the HDGCP as it builds an innovative global network for research in the human sciences.

These themes will be added to, and likely modified, as planning proceeds. National and regional projects will also become part of the Program, as it develops into the 1990s.

IAN BURTON and PETER TIMMERMAN

CHAPTER 11

# People Pressure

## SUSAN A. McDANIEL

People. Seldom thought of like other resources, people may be our most valuable resource. People provide labour, without which we could not survive. Even more important, it's people with ideas who can solve the problems we face now and in the future. But people also use resources wastefully and destroy forest, lakes, atmospheres, and habitats by their ways of living as well as by their sheer numbers. As a global challenge, people and the growth of human populations are a paradox, both threatening life on this planet and having in their power the ability to retard or prevent the worst effects of global change.

It's hard, maybe impossible, to talk about population in a global perspective, since regions of the world differ so dramatically in their population problems. In the so-called developing (sometimes called Third World) countries there are young, quickly growing populations. In more developed parts, low birth-rates and aging populations present a different challenge. Between these two extremes there is no longer any middle ground.

Viewed another way, however, all population problems are global, because we are all so intimately connected on the planet Earth. Immigration (including refugee flows), wars, regional disputes and unrest, famine and pestilence, changing climatic conditions, emergency and continuing aid, trade and markets, and national debts—all are global challenges, since none of these issues are confined to the boundaries of a single political state. Population issues are intertwined with global challenges, but are neither direct causes nor direct consequences. The system has become so complex that to think in terms of simple cause and effect is no longer appropriate. Causes have become effects too, as forces feed back on each other.

Fortunately the pace of world population growth, which reached the highest rate ever in 1955 and 1965, has been slowing down since.

*By the year 2000 the population of the world will be six billion. (World Bank Photo.)*

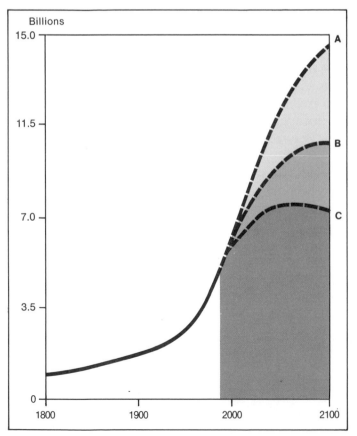

FIGURE 62. *Three world population projections from the United Nations (A, B, and C). As survival rates have improved with public health, the doubling period for world population has shortened from centuries to decades. All three different projections assume that birth-rates will fall to zero growth in the next fifty years, but with different rates of change. (UN Demographic Projections.)*

However, global population today is still increasing fast, currently by one billion every eleven years (Figure 62). Best predictions now are that the rate will be 1.5 per cent a year by 2000—when the world population will be six billion. Although the rate of increase is slowing globally, it is slowing at a faster pace in western countries. By 2000, for example, China and India alone could have a combined population of nearly two billion, only slightly less than the population of the entire world in 1950!

Our optimism about reduced world population growth is therefore tinged with concern. Nearly ninety per cent of the net increase from 1965 to 1985 is in developing countries. At the same time, developed countries are experiencing the lowest birth-rates ever recorded. Many Western European countries, as well as the United States and Canada, are worried about aging populations, actual population declines, and potential labour shortages. But developing countries, which are potentially large suppliers of labour, are wasting lives through illness and disability, premature death, and lack of opportunity. Gaps between the richer and poorer part of the world widen. We are becoming a demographically divided globe. (Figure 63.)

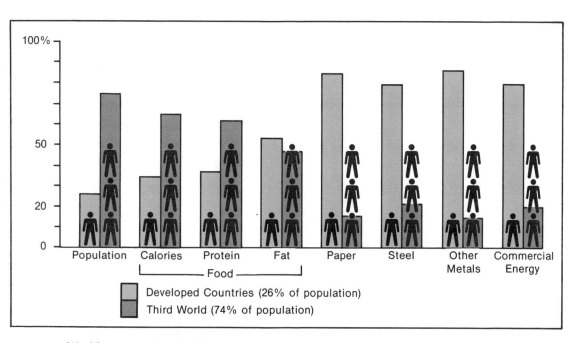

FIGURE 63. *World consumption distribution. Developing countries, with about three quarters of the population of the world, consume much less than the quarter of the world in developed countries. (WCED estimates based on data from FAO, UN Statistical Office, UNCTD, and the American Metal Association.)*

*We are becoming a demographically divided globe, and gaps between richer and poorer parts of the world are widening. These children are being fed at a relief camp in Ethiopia. (World Bank Photo.)*

## INTERCONNECTIONS OF POPULATION
## AND OTHER GLOBAL CHALLENGES

Population change is one of the remarkable features of our contemporary world. Like a flood with crests and lulls, it is driven by forces little understood and thus not easy to predict. But population growth does not occur in a vacuum; it is linked to environmental forces, to economic conditions, to culture, and to many other factors. Nevertheless, experts and non-experts alike find it hard to resist the temptation to view population as the driving force behind resource depletion, pollution, lack of economic growth, and a range of other problems.

It is true that more people tend to use more resources—from arable land to natural gas—but the relationship of population to resources is not direct, nor is it simple. Between population and resources are vital mediators such as technology and social organization. For example, it has been argued that if in India agriculture were as intensive as Japanese agriculture, and as technologically advanced in the use of terracing, irrigation, chemical fertilizers and weed/pest controls, it would support more than twice its present population with improved nutrition. This argument is silly because, although it considers technology, it overlooks the fundamental socio-historical reasons why Indian agriculture is less intensive than Japanese. Without attention to these crucial factors, this argument remains hypothetical and useless as a means of improving the situation. One more example. The population of Bangladesh, among the most dense in the world and growing rapidly, uses far fewer resources per capita and in total than does the population, one quarter the size, of Canada. Both differing levels of technological development and social organization, as well as resource availability, account for this.

In the same way population increase cannot be blamed directly for environmental problems. While more people tend to add more pollution, the potential for environmental disasters are greatest in those countries with the highest levels of industrial development, rather than those with the densest populations. Witness the nuclear-power disasters at Chernobyl in the USSR and Three Mile Island in the US, and the fire involving cancer-causing PCBs at St Basile LeGrand in Quebec. Similarly, acid rain is a product of poor industrial practices rather than over-population.

Population growth must be seen as one factor that is pressuring resources and the environment, but it is not the only factor, and perhaps not the most important. Slowed population growth, although offering the possibility of ameliorating many problems, will not reduce destruction of environment and resources; it will only lower the rate of increase. Also, the effects of slowed population growth vary from region to region, even though the population-resource balance is ultimately global. For example, in more labour-intensive economies, more people could be an asset to economic growth. Paradoxically, reduced population growth poses its own challenges: an aging population in developed countries overloads pension and health-care services while the work-age population dwindles.

Nor is over-population only, or even predominantly, a Third World problem. Many people who have seen it that way thought the problem would be easily solved if couples in the Third World would reduce family size. About twenty years ago I, like others, thought that

## Population profiles
Proportion of men and women by age 1990

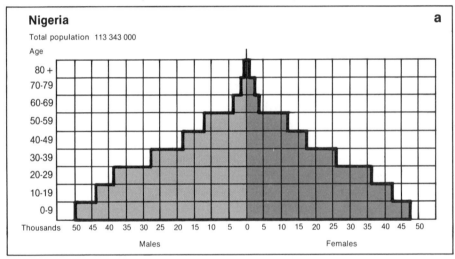

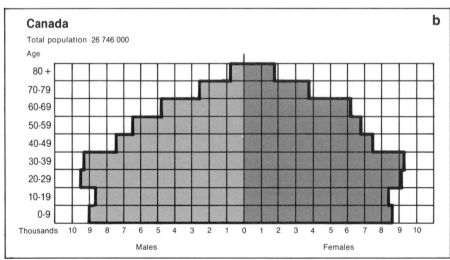

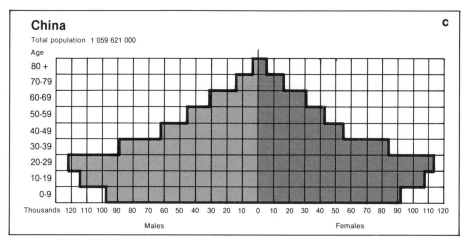

FIGURE 64. *Population profiles estimated for 1990, showing different age structures for Nigeria (a), Canada (b), and China (c). (UN World Demographic Estimates and Projections, 1988.)*

resources not spent on additional children would go to economic growth and advancement. Our simple-minded models failed to recognize that hardly any resources were spent on the children who lived in dire poverty in most Third World situations, and that, contrary to our Western preconceptions, children themselves were actually resources—the hope for the future, possible lottery tickets to a better life. We now know very well that each person born in the developed world inevitably does more environmental damage, in terms of both resources consumed and waste produced and in capacity to destroy the environment, than a person born in the Third World. We have also learned from our own history that declining birth-rates tend to accompany economic development, not precede it. Further, we have learned that externally imported attempts aimed at reducing numbers of births, no matter how sophisticated or well intended, were pretty well doomed to failure. This was not because Third World couples were completely committed to having large families, nor even that family-planning programs were badly designed (although many were), but rather that family size is bound up with culture, social organization, level of development, and multiple other not-easily-changed aspects of life. (Figure 64.) Changing one factor in isolation from others simply does not work.

Population and other global challenges are intertwined because of the world economic system. Sky-rocketing debts encouraged by developed countries in many Third World areas have now reached such a magnitude that, if reneged on, they could produce economic recessions in the developed world. Third World countries provide

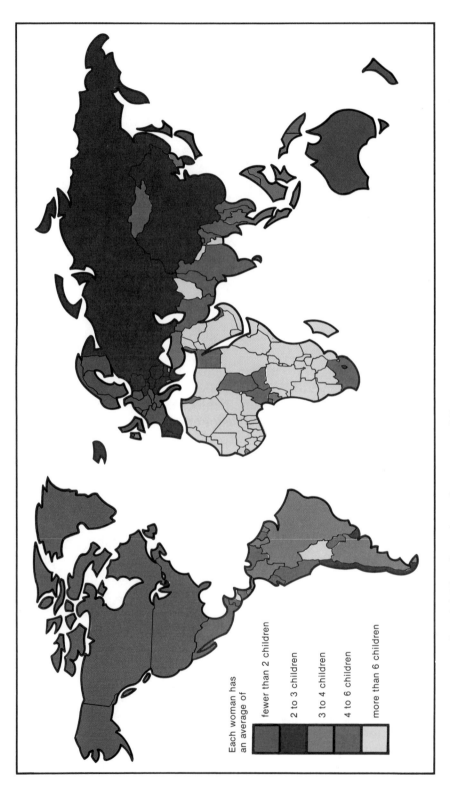

FIGURE 65. *Average number of children borne by each woman in different countries. (After The New Internationalist, March 1989.)*

markets, despite their poverty, for developed countries selling everything from Coca-Cola to cigarettes to agricultural equipment. Labour goes to developed countries from the Third World in the form of qualified immigrants (the familiar 'brain drain'), guest or migratory workers, or refugees. Aid, in dollars and in kind, links regions. These many connections are far from uniform, nor are they even working toward the same ends. For example, the economic and political positions of today's superpowers could be seriously threatened by the economic advancement of China, and yet industry in both the United States and the USSR stands to benefit from the opening up of new markets in that large country. Successful family-planning programs may not be entirely welcomed by interests concerned about competition from rapidly developing countries. If they should be successful, resources previously spent on burgeoning populations could now be redirected to economic developments which, combined with a cheap and available labour force, could make Third World economies competitive on the world market. Witness South Korea's recent rise as an industrial nation: General Motors must now compete with the much-cheaper-to-make—because Korean labour is cheap—Hyundai car.

## WHAT DO WE KNOW NOW?

The answer to this question is both simple and complex. We certainly know more than we knew previously, but part of our new knowledge includes the recognition of how little we do know. As revealed elsewhere in this book, we know a lot about the physical dimensions of our global challenges. But all of them include a human dimension, and that is far less well understood. In the case of population, the way people act is the most important component.

Population change is at once a biological and a social process. Pregnancy and birth are biological and the maximum number of possible births has a biological limit, but people everywhere keep reproduction below this limit. In many societies, such as our own Western society, the constraints are direct and include widespread use of contraception, heavy and growing reliance on sterilization, and the limited availability of abortion. In many societies, also including our own, social constraints are less directly associated with birth prevention, but they are effective just the same. Societal rules, laws, or religious norms govern when couples can marry or whether births outside of marriage will be tolerated. Other rules determine who can marry whom, or whether widows should remarry. All these rules affect risk of pregnancy and therefore birth-rates.

The social roles women play are fundamentally important to birth and death levels. If women's status is closely bound to being wives and mothers and producing heirs (typically males), and if women are denied independent choices, birth-rates tend to be generally high and very difficult to reduce. Similarly, women as the 'front-line' health caregivers can do a great deal to prevent premature death (especially of children) and to share their knowledge. We now know without question that providing new opportunities for women—including increased education, increased control over finances, increased status at home, as well as increased control over their own bodies—all bring about lower fertility. (Figure 65.)

*Women in Bangladesh listening to a radio program on family planning. Fertility rates are directly linked to the education and status of women. (World Bank Photo.)*

We also know how to effect demographic change, although we seldom act consistently on this knowledge. Birth-rates and death-rates tend to fall as literacy, particularly women's literacy, increases (Figure 66). Both rates respond, although not equally, to health services and public-health improvements. By far the most important force in demographic change is the reduction of poverty, which can be done with surprisingly limited investment of resources. The World Bank, for example, estimates that malnutrition could be eliminated with only two or three per cent of world grains being redistributed to people in hunger. Reducing poverty also improves people's health, energy levels, and quality of life. Similarly, effective health services could be provided in most Third World countries today for the same costs of their present ineffective, inefficient, and often inappropriate health systems. It would require a reorientation of priorities to develop effective primary health care for rural, and the largely disenfranchised burgeoning urban, populations of the Third World. It would also need a redistribution of economic, health, and social services from developed to developing parts of the world.

We know that the average North American male could add a decade to his life expectancy if he stopped smoking, drinking, eating too much of the wrong foods, and getting too little exercise. These are problems of excess, of affluence, of abundance. If money spent on excesses could be redirected to Third World scarcities, all of us would be healthier, and likely happier. Increased health, and the hope that accompanies it, generally leads, as it did in our own history, to greater economic and social productivity and ultimately to reduced population growth.

We also know that community-based programs of development, in conjunction with the provision of health and family-planning services, tend to be more effective than either initiative taken separately, espe-

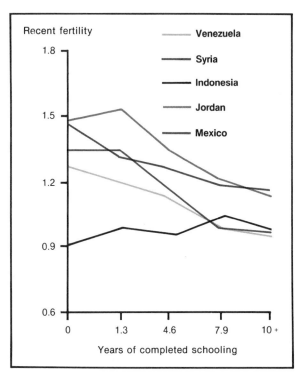

FIGURE 66. *As the education of women increases, the number of children born per woman decreases. (UN World Population Trends, 1983.)*

cially if local people are involved in conducting them in their own ways, consistent with their own vision of their past and future. Recent debates among population experts about whether family planning or development is the most effective approach to global population challenges are futile and misleading. Both are necessary, and may be inseparable—acting locally but thinking globally.

Lastly, in a far-from-complete summary, we also know that learning and teaching are not one-way streets. Learners often teach more than they learn. In the arrogance of a few decades ago (which persists in some circles), we in the developed world thought we had answers to problems faced by our Third World cousins. We eventually learned, but not quickly enough, that there is more than one way to be 'advanced'. What we in the Western middle class might possess in terms of modern conveniences and material wealth, we lack in terms of human contentment and cultural/social wealth. For many of us it was our children who rejected our values in favour of what they considered more important—it took the shock of that to convince us of the incompleteness of our society and its values (see Chapter 10).

There is no doubt that our knowledge about the population challenge is wide, solid, and growing. However, it is quite another matter to act on this knowledge to bring about change. There are many obstacles: vested economic interests, unwillingness to accept the truth of what we know and lack of courage to apply it, fear of social change, and more. In many ways the present time is a time of parentheses, a time between eras, in which we cling to the past and the present even as we watch the present slip away into the future.

## THE GLOBAL POSSIBLE AND THE GLOBAL PROBABLE

We do not know the ultimate carrying capacity of our globe, mediated as it is through social organization. Uncertainty raises our collective anxiety, as if we are travelling at high speed down a highway with little idea of where we might end up—if in fact we end up anywhere but in a calamitous collision. Psychologists say that uncertainty and anxiety together make us vulnerable either to unfettered optimism (things are so bad that they can only get better) or to dire pessimism (things can only get much worse). Experts are not immune to one or the other tendency either.

Speculations about the future of population growth and its problems seem particularly prone to over-optimism and over-pessimism. This may be because we actually know so little about the bigger social forces involved in having children, dying, moving from place to place. But also these basic population processes are related to factors we do understand, such as sexual expression, disease and sanitation, food requirements, droughts, and other physical or biological forces. For this reason, and because population processes can be quantified, experts tend to be deceived. For example, one model known as 'the sunlight model' bases estimates of future population on how much space each human requires in terms of kilocalories of sunlight reaching earth (it turns out that a person could live in one square metre). By this model the so-called carrying capacity of the earth is 26,000 times the 5 billion population in the world today! Overlooked in this absurd model, of course, is the complexity of physical, socio-logical, cultural, and historical forces that govern human settlement and determine human needs. For example, the consumption of energy in every respect is much greater per person in the United States than it is in Nepal, for a number of reasons. The sunlight model suffers from simplistic over-optimism.

The globe does not have an infinitely expandable carrying capacity. People live in places such as the barrios of Tegucigalpa, Honduras, or the slums of Calcutta, in such great numbers that the carrying capacity of these environments has long ago been exceeded. Even in these dire circumstances, however, much could be achieved by human caring, in the form of adequate sanitation, fresh water, and garbage disposal. Not only would these improvements add greatly to the quality of life,

*The globe does not have an infinitely expandable carrying capacity. That of slum environments like those of Manila have long been exceeded. (N. McKee, IDRC Photo.)*

they would enable these unfortunate people to acquire the energy to contribute to a better community for themselves, and consequently for us all.

To think about the future is to contemplate change, to imagine what is possible. This is imperative. But the future grows out of both the past and the present. This continuity need not mean fearing the future, or change, so that only the past looks appealing, as some contemporary politicians and analysts like to think. Given the constant tension between change and continuity, what can we know about the future of population as a global challenge? Certainly the farther ahead we project, the more uncertain we are. At the same time, the nature and size of our global challenges require us to think imaginatively and creatively about what is happening now. Among current trends, which are long-term and which short-lived? Are changes occurring now that indicate what is to come?

In contemplating these big questions, we should keep some thoughts in mind. First, changes can and do occur at different levels and in different ways. For example, climatic changes in Saharan Africa may have more long-term effects on population in those regions than any kind of development or family-planning program could. Second, changes have different time-frames. The transition from high birth-rates and death-rates was slow in the West. This may not be true in underdeveloped countries. Third, and very important, different perceptions of problems can be equally valid. Anthropologists have long recognized the wisdom of different perspectives. Population experts are coming to this realization slowly. Fourth, even in a demo-graphically divided world, it is possible for local actions to have global possibilities—a successful development program in one region, for example, can free resources for use elsewhere.

The 'global possible' involves shifting emphases in several ways. It no longer serves our interests, nor those of others on our globe, to be condescending and pious about what we in the developed world have to 'offer'. The wholesale exporting of Western culture and values has done little good. Even as we become less pious, our values shift to less hierarchical and rigid patterns. The contextual and caring thought-patterns that are generally more characteristic of women might prove to be a good basis for solving global challenges, rather than the more competitive win/lose patterns of men. In 'helping' solve Third World problems by giving women more opportunities, we may be called upon to do the same at home. Education and meaningful paid work for women of the Third World will go a long way towards solving problems arising from birth-rates outstripping resources. In recog-nizing the centrality of women in the solution of global challenges in the Third World, we may be forced to do the same in Canada, the US,

Europe, and the USSR. Women's social roles and ways of caring may be crucial elements.

This is also a dimension to the 'global possible' that strikes close to home—another component of acting locally but thinking globally. We need a creative readjustment in Western values just as much as in Third World values. There is little doubt that we live in a dangerous era. We must recognize that danger if we are to be realistically optimistic. New ways of thinking can bear the seed of change, and they must include science and social science. Examples of an awakening to the need for new thinking are found in Europe, where a political party devoted to environmental concerns (the Greens) has gained popular support, in the continued popularity of health-food stores in Canada and the US, in protests against military installations in England, and in the growing number of people who are beginning to say to our political leaders that people and social issues matter more than profits. The transformation of today's global challenges, including (or maybe even particularly) the population challenge, into the global possible is likely only if we believe it can happen, and will work to make it happen.

# The Demography of Development

## THE INCREASING NUMBERS OF POOR

The present and prospective pattern of birth-rates increases the number of poor in the world and lowers world-average income. Since birth-rates and natural increase are greatest in the poorest countries, the weight of population is shifting towards those countries. Suppose per-capita incomes of $500 in poor countries and $9,000 in rich countries, and rates of increase of 2 per cent and 0.5 per cent respectively, along with populations of 3.8 billion and 1.2 billion. (For simplicity, suppose no change in income per capita in individual countries.) If, in 1987, the world

| YEAR | POPULATION | | | INCOME |
|------|------------|--|--|--------|
| | Developed Countries | | | |
| | Least | | Most | Per capita |
| 1987 | 3.800 | Billion | 1.200 | $2540 |
| 1988 | 3.876 | | 1.206 | 2517 |
| 1989 | 3.954 | | 1.212 | 2494 |
| 1990 | 4.033 | | 1.218 | 2472 |
| 1995 | 4.452 | | 1.249 | 2362 |
| 2000 | 4.916 | | 1.280 | 2256 |

average income was $2,540, then because the poor are increasing faster than the rich, the following year it will be $2,517, and so continue downward. This scale is illustrated on page 239.

We hope that both in the least-developed and the developed countries income per head would increase, but even if such a desirable change takes place within individual countries, this downward tendency would be superimposed owing to shifting proportions.

## WORKING-AGE POPULATION

Progress from high to low birth- and death-rates brings a more mature population, with fewer children and more old people in

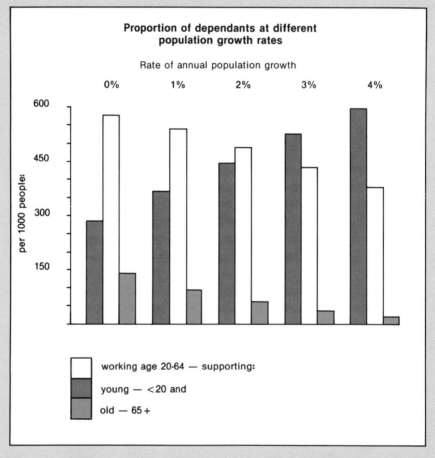

FIGURE 67. *Different rates of population increase, and their effect on the proportion of dependants. The progress from high to low birth and death rates leads to a more mature population, with fewer children and more old people.*

proportion. People have always wanted a long life, and with the advance of sanitation and medicine, more and more can satisfy that wish. But if we live longer, then the community is bound to have more old people. What is important is that total dependency—of those too young and those too old to work, taken as a proportion of the labour force—goes down sharply with the slowing of population growth. One of the salutary results of family planning is a population of which the larger proportion is of working age.

Figure 67 compares different rates of increase, for populations increasing at 0, 1, 2, 3, and 4 per cent a year, all supposing a life-expectancy of 65 years—about the world average expected over the next decade or two. The proportion of working-age people goes steadily up from over a third to over a half as the rate of growth slows from four per cent to zero. On the average, each person aged 20 to 64 supports 1.644 young and old people when the population is increasing at four per cent per year, and supports less than one person (0.736) when it is stationary.

## EDUCATION AND UNEMPLOYMENT

Because high priority has been given to education in many developing countries, unemployment takes a particular form when the newly educated are unable to find work that makes use of their training. The movement of the educated to cities where there is not the capital to employ them effectively creates disruptive political pressures.

Lack of jobs in the developing countries by no means signifies that there is no work to be done. The roads need repairs, the streets need cleaning, the rubber trees need replanting, the fields need weeding. People are standing around doing little or nothing; joining them to the necessary jobs is an obvious necessity. Yet underdevelopment consists partly in the fact that the work and the people are not brought together. The needed factor is the organization that would put the parts together; this applies in all developing countries, whether they are market or command economies. The amount of such organization that comes into view does not seem proportional to population.

NATHAN KEYFITZ

# Family Planning in Zimbabwe

Forty per cent of married women of child-bearing age in Zimbabwe use some form of contraceptive—four time as many as in other African countries south of the Sahara Desert. Many more know about contraception methods and where they can obtain them. Choice and education are the reasons for the remarkable success of the planning program of this Southeast African country with a population of about eight million.

Family-planning services were first established in Zimbabwe (then Rhodesia) in 1953, and the Family Planning Association was formed in 1965. However, these early programs featured little community involvement. People were given contraceptives they did not request and knew little about, and some were even sterilized without consent. The white-minority regime's family-planning program was reputedly anti-black.

In 1981 the government of the newly independent Zimbabwe took control of the Association. The Zimbabwe National Family Planning Council (ZNFPC) emerged in 1984, when activities were reorganized under the Ministry of Health. Now a staff of over 1,000, and thirty-five family-planning clinics, give Zimbabweans access to condoms and pills through a community-based distribution program. Community family-planning staff members are chosen by the villagers themselves, and local distribution centres make it possible for rural women to reduce travel to an average of 36 minutes for contraceptives; urban women, 15 minutes. Contraceptives are free to those with a low income.

Zimbabwe's success in family planning—illustrated by a fertility-rate decline from 7 per woman in 1980 to 5.6 estimated for 1988—is attributed to government commitment to its 1986-90 five-year plan to reduce population growth. (Even television soap operas promote family planning.) The relatively high standard of living and education—more women of child-bearing age are in secondary school or the work force—are also factors. The goal is that by 1990 seventy per cent will be using contraceptives.

Sub-Saharan countries experience many stresses: large, fast-growing populations; food and supply shortages; soil erosion and loss of vegetation. Despite the high growth-rate, many countries have no family-planning programs—Zimbabwe is an anomaly and an example of what might be done. But even Zimbabwe's population will double in seventeen years at its current growth rate of 3.6 per cent. The population growth-rate for all of Africa is 2.6 per cent (for Canada it is 1.1 per cent). There are twenty-two people per square kilometre in Zimbabwe, compared to three in Canada.

## KUBATSIRANA PILOT PROJECT

Kubatsirana means self-help, the name of an innovative joint pilot-project by the ZNFPC and the Adult Literacy Organization of Zimbabwe. Begun in 1986 in seventeen communities in the Goromonzi district 70 kilometres east of Harare, the three-part program was based on the three needs women themselves identi-fied: income generation, literacy, and family planning and health. An underlying principle was that people should have sufficient education in family-planning matters to be able to make their own choices. In each community one member was chosen to be trained as a Field Co-ordinator, to help bridge communica-tion or cultural gaps between rural participants and the city-dwelling organizers. One other person was also chosen by the women in each village to be trained as a 'community-based teacher'. Teachers were paid during training, which took place locally and included financial management and on-site practice. In turn they trained new teachers.

Community distribution workers, also local women, were able to reach between one and two thousand women each month and to make periodic repeat visits for as long as six months, to make sure the chosen means of family planning were working.

## MALE MOTIVATION

A 1984 reproductive health survey revealed that 42 out of every hundred Zimbabwean women thought that men alone should be making family-planning decisions. (Eight thought the man and woman should decide together; 36 thought the woman alone should decide.) The ZNFPC recognized the need for a program to change the way men view family planning: to increase their knowledge of methods and to encourage their participation. Radio and press advertisements, posters and pamphlets were spread nationwide. Men were taken on tours of family-planning centres and discussed the benefits of having small families.

The ZNFPC distributed 2.7 million cycles of birth-control pills in 1986—thirty per cent more than in 1985. The use of barrier methods of birth control and intra-uterine devices also increased significantly.

CHAPTER 12

# From Technological Fix
# to Appropriate Technology

## DONNA SMYTH

Think of a wind-surfer riding the waves, displaying poise and balance, strength and control. We would all like to ride the waves of change in this way—working with Nature, not against her, using our human flexibility, along with tools, so that we, like the child at play, discover the infinite possibilities of Being. The root of 'technology' is the Greek word *tekhne*, which means art, craft, or skill. Our Western culture began in a society where the technician was an integrated human being—artisan, artist, and scientist; where technologies served human needs.

Now we find ourselves in the midst of what some call a technological 'revolution'. New technologies—especially military, communication, and biotechnological—are being developed and applied at a faster rate than ever before. Some analysts claim that these technologies are being developed for their own sake and have outrun human control, that we are in the grip of technological imperatives. Others claim that the new technologies merely open doors, and that we have a choice as to whether or not we walk through. The entirely optimistic, which include many Western policy-makers, claim that 'high technology' will save national economies, liberate citizens from the drudgery of work, and eventually create a universal global culture—a kind of Utopia, in fact. The most pessimistic, on the other hand, claim that technology will lead either to global destruction or to the creation of a 'brave new world' of control and surveillance of workers and citizens by an élite class of scientists and technocrats funded by global commercial interests.

I think of myself as a technological realist, preferring to consider technologies within the context of social and natural ecosystems where they are conceived, developed, and applied. So considered, technologies are never neutral—they are not simply tools we can put to good or bad use. They reflect our cultural values. In Western nations, for example, both science and technology are traditionally viewed as a means to control and dominate Nature. Practitioners, forgetting that humans are both a part of and dependent on Nature for survival, 'displaced' themselves to take on the role of scientific observers and manipulators, believing that the mess created by one technology could be solved by another, by a technological fix. Until the present global crisis, this faith allowed us to largely ignore the human and environmental consequences of our technological choices.

It also obscured the extent to which we are shaped by the technologies we create. I would even argue that technology has altered the nature of human consciousness, bringing about a kind of global consciousness that has both creative and profoundly destructive aspects. One symbol of this mind-change is the image, beamed back to Earth by the American astronauts in the 1960s, of our lovely planet suspended in space. It signalled our global interdependence and vulnerability: what happens on our part of the planet will have some kind of effect, at some time, on all the other parts. I call that perception global consciousness.

At the same time, military technology has presented us with a grim message. The ultimate control of Nature is the power to destroy her and ourselves absolutely. Living under the nuclear shadow forces many of us to conclude that war is obsolete. Any use of nuclear weapons can escalate into a global conflict that no nation can 'win'. And quite apart from nuclear weapons, new chemical and biological weapons can extend the range of environmental and human damage in regional and potentially global conflicts. It's the same with the new conventional weapons. Computerized 'smart' guns and missiles are being rapidly distributed throughout the world, particularly in the less-developed countries whose governments, since 1984, are the major customers of the global-arms bazaar (Figure 68). As regional conflicts, like the Iran-Iraqi War, become more deadly for the participants, they become more dangerous for the rest of the world.

The transfer of military technology, both artifacts and 'know-how', from richer to poorer nations demonstrates, on a global scale, how technologies are always linked to existing power structures and may have economic and social consequences beyond their immediate use (Figure 69). The consequences are both local and global. In poorer nations, children starve while money that could buy food buys weapons. In the richer nations, military budgets escalate as the costs of new and sophisticated weapons soar—for example, the Stealth technology recently developed in the US costs, at a conservative estimate, $500 million per plane. The immediate result? Social programs,

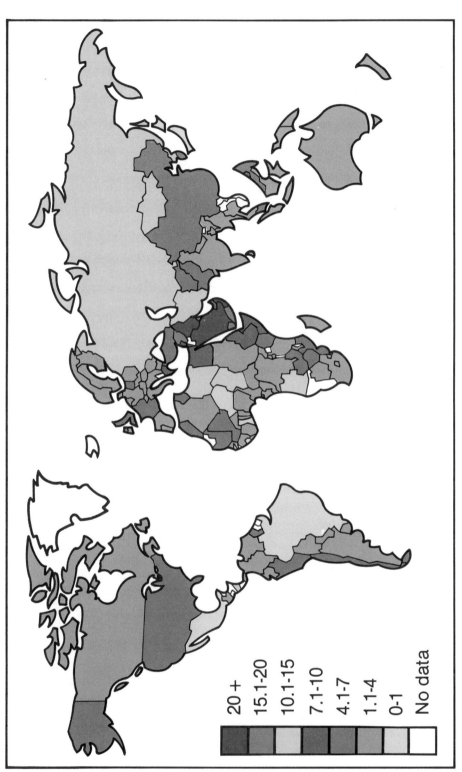

FIGURE 68. *Military expenditure as percentage of Gross National Product, 1984. Less-developed countries are the major consumers at the global arms bazaar. This map does not distinguish the really high purchasers: Iraq with 50 per cent of the GNP, Oman with 27.7 per cent, and Israel with 27.1 per cent. Figures were not available for Cambodia, Laos, and Vietnam. (Ruth Leger Sivard, World Military and So-cial Expenditures, 1987-88.)*

Legend:

20 +
15.1-20
10.1-15
7.1-10
4.1-7
1.1-4
0-1
No data

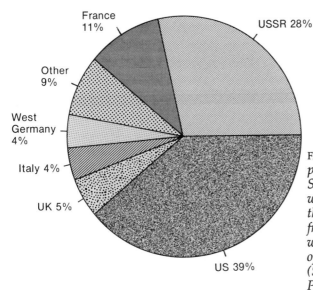

FIGURE 69. *Major arms exporters, 1981-5. The United States and the USSR together were responsible for two-thirds of the arms exported from 1981 to 1985. Most went to developing nations, often in subsidized purchases. (Stockholm International Peace Research Institute Yearbook, 1986.)*

education, health care, environmental-care and other non-military job creation, scientific and technological research and development—all are cut.

## POLLUTION

Another aspect of global consciousness linked to technology is our increased awareness of pollution (see Chapters 2 and 6). During the 1950s chemical and nuclear industries, including the nuclear-weapons industry, developed and expanded rapidly. Pesticides like DDT were marketed as indispensable to agriculture and, indeed, to modern life; housewives were encouraged to use it to control house-flies. Short-term benefits obscured long-term risks. Rachel Carson's *Silent Spring* warned us in 1962 that along with pests, beneficial insects and birds were being destroyed and ecological systems were being degraded. DDT residuals were even found in human fatty tissue.

Global air- and water-flows mean that no nation or person is safe. DDT was eventually banned in Western industrialized nations, but not in developing nations, and traces are still found in polar bear fat in the Arctic, for instance. When DDT was banned, other pesticides took its place. Today thousands of new chemicals appear on the market each year. Toxic-waste disposal is not going to go away.

Similarly, radioactive waste accumulation in nations that mine and mill uranium ore has created 'dumps' potentially dangerous to human health and the ecosphere for thousands of years. Some experts say

*Insecticide sprays that have been banned in Western industrialized nations are still used in developing nations. (World Bank Photo.)*

there is still no safe way to dispose of nuclear waste; others pin their hopes on long-term storage in underground caverns.

In the 1980s we have to admit that the creation of general waste is leaping far ahead of the technology to dispose of it. For some problems a technical 'fix' is possible—at a cost—and many technologies are still in the developmental stages. No nation wants to become the world's dumping ground. The richer nations have used developing countries like Nigeria as dump sites, but newly industrialized nations are tightening their regulations and have their own pollution and waste problems.

International media coverage of tragic accidents—among them the 1984 methyl isocyanate leak at the Union Carbide pesticide plant in Bhopal, India, and the 1987 Chernobyl nuclear-reactor accident in the Soviet Union—have heightened our awareness of the problems. And the international scientific community and non-governmental environmental groups have pointed up our own collusion in injecting a steady flow of impurities into the environment: we drive cars, have refrigerators, burn fossil fuels—contaminating the atmosphere with ecologically damaging ingredients.

## COMMUNICATIONS TECHNOLOGY

One of the necessary conditions for global consciousness has been electronic-communications technology; the Live-Aid TV concerts for African famine relief in 1985 reached an audience of 1.5 billion in 169 countries. Because radio and TV can speak to print-illiterate audiences, their potential is exciting. I view the electronic-oriented culture that has developed over the past twenty years as complementary to the print-oriented culture of the industrial nations and essential to communication with developing or newly industrialized nations. So far, however, the international diffusion of communications technology has produced an asymmetrical pattern: while most poorer nations have developed their own radio networks, television hardware and programming tend to be developed and distributed only by the rich nations, with the US taking the lead. Children in Peru end up watching American cartoon shows; the Inuit in Canada's North watch American soap operas.

This particular cultural drift towards homogenization reflects the world's economic power structure. Some optimists call the phenomenon 'convergence', part of a new global culture. Yet the various

*Farmers gather each morning around a radio to listen to an agricultural broadcast written at the agricultural college at Lyallpur, Pakistan. (World Bank Photo.)*

separatist and nationalist movements around the world suggest that we still need local roots, a place and a way of life to validate individual identity. How can we be global citizens and yet preserve indigenous, regional, and national cultures? A truly global culture must reflect diversity, multiplicity, and plurality as well as what we have in common. And this must mean more than television entertainment or a commercial culture represented by a McDonald's in Moscow, a Kentucky Fried Chicken outlet in Beijing.

Technology gives us the means to communicate globally, but the way we do it depends on non-technical choices. The potential of a particular technology depends on its application in a particular economic-social-political field, which dictates how, when, and where it will be used. A technology, or a technological system like nuclear energy, may be developed but not fully implemented or exploited if politics and economics say 'no'. Conversely, economics or politics may push a development like computers.

In the 1980s these choices are complicated by a cluster of new technologies that, in interrelation with communications technologies, are transforming our collective knowledge and our lives.

## CULTURAL REVOLUTION

Are these new technologies, including satellite-based global data-gathering systems and computers, causing a cultural revolution? I, among others, would say yes: we are entering the Information Age. In the richer nations these technologies have already penetrated most of our living through personal computers and automated banking, telephone exchanges, offices, libraries, and workplaces. Most of these changes have occurred within an incredibly short time—the last ten years—and the rate of innovation and implementation increases. The traditional long lead-time in developing new technologies, the 'lag time', has shrunk to virtually zero in some areas.

Part of the impetus comes from the computer industry; their products are faster, more efficient, and they have shrunk in size and cost since 1946, when the first electronic computer came on line. Christopher Rowe, a British technological analyst, in his 1986 book *People and Chips*, says: 'If the motor industry had developed at the same speed as the computer industry, then a Rolls Royce today would cost around three pounds, it would do three million miles to the gallon, its speed would be staggering, and one could put five on a fingertip.'

The obvious historical comparison is the Industrial Revolution of the late-eighteenth and early-nineteenth centuries (Figure 70), which produced a cluster of interrelated changes, from new technologies to

# Three Industrial Revolutions

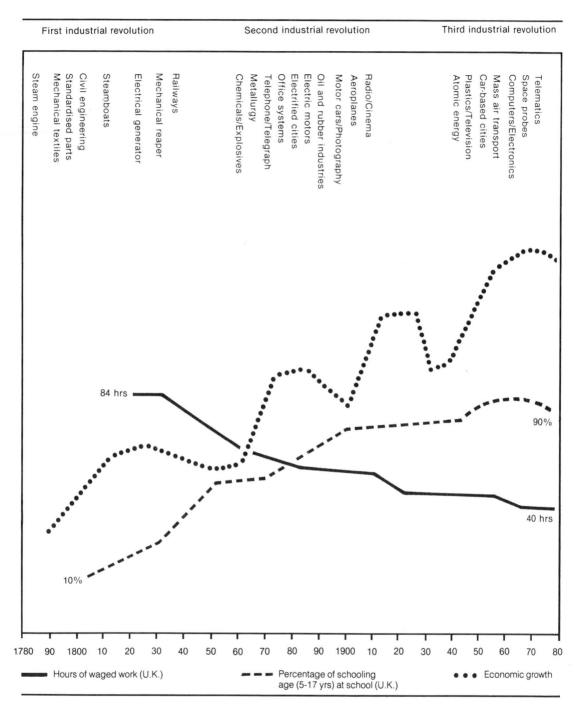

FIGURE 70. *The heavy dotted line indicates increasing economic growth in the world during three phases of industrial development. Unlike the line for education and hours of waged work, the progress is not smooth. Rapid technological change may actually disrupt economies, which then make social and economic adjustments, leading to further more rapid growth. After* The New Internationalist, *August 1986.*

political-economic-social patterns. Key inventions like the steam engine resulted in new production processes. In turn these generated new production units and work patterns; for example, the factory influenced movement from country to city and resulted in the dislocation and 'deskilling' of thousands of men, women, and children. Capital resources were used to intensify mechanization and automation and eventually created new classes: worker and boss, proletariat and bourgeoisie.

When we look back, we see how embedding new technologies caused social disruption and suffering. At a certain stage in these radical cultural changes, ordinary people were driven by technological-economic demands. In the nineteenth century this was called Progress. Those who could adapt survived, those who could not disappeared. The survivors were eventually 'reskilled' and 'relocated', but the suffering and poverty involved was beyond cost-accounting.

Obviously the scale of a particular technology or technological system, its own rate of change and application and its interaction with the rest of society, all have to be taken into account when we try to assess the Information Age. In the richer nations, computerization and robotization transform traditional manufacturing and service industries—as in the automobile industry, where automation is deemed more cost-effective than human labour. Indeed, in many work-places the 'human operating unit' has become or will become more or less redundant. Jobs are lost; the automated office is transforming clerical work usually done by women.

At the same time, as global markets open up and capital is freed of its traditional dependence on labour, we observe the restructuring of the international economic order. During the 1970s transnational corporations decentralized. Older industrial centres declined while new plants went up in poorer countries with cheap labour and fewer regulations. Atari, the videogame/home-computer giant—with plants in Singapore, Puerto Rico, and Ireland—laid off 1,700 American workers in 1983 and moved its Silicon Valley production to Hong Kong and Taiwan. Newly industrialized nations like South Korea leapfrogged over traditional industrial development to establish an electronics/automated industrial base competitive in the global market. While Japan and the US vie for dominance in the high-tech race for global markets, less-developed countries, particularly in Africa, can hardly begin to compete.

The development and implementation of complex, sophisticated, and very expensive new technologies may exacerbate existing inequalities, or set up new ones between rich and poor nations. The high-tech industries themselves create a divided work force. In Silicon Valley companies, for instance, engineers and computer scientists are at one end of the pay and skill scale; at the other are low-wage, low-skill workers, many of them women and/or racial or ethnic minority-group members.

## The Information Age

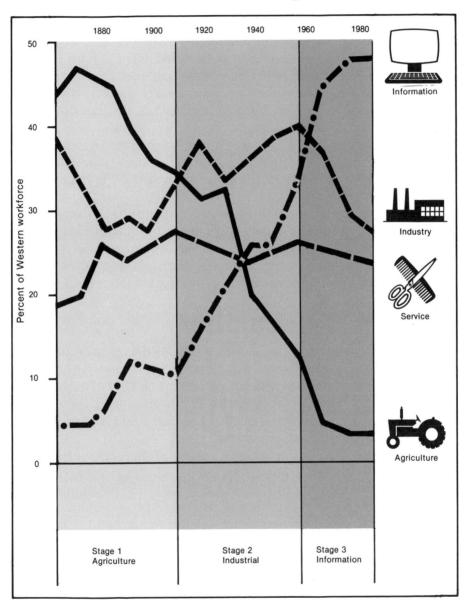

FIGURE 71. *An increasing percentage of the work-force in most Western countries are now handlers of information and providers of physical goods. Adapted from* The New Internationalist, *August 1986.*

Optimists point to the growth of the information sector of industrialized economies. They predict that by the year 2000 two-thirds of all work will be devoted to the gathering and dissemination of information (Figure 71). Computer-literacy will be as vital as print-literacy, and workers will have to learn the new technologies as they evolve. In the professional classes this has already happened, but in uneven patterns—patches, as it were—depending on profession and nationality. Within the international scientific community, satellite data-gathering and remote-sensing techniques have become an integral part of global-change research. To understand what is happening to global ecosystems we need to map solar radiation, monitor atmosphere, oceans, vegetation, sea ice, permafrost, glaciers, rivers, and lakes. This is one area where Canada, a middle power, has developed expertise. Since 1972 the Canada Centre for Remote Sensing has received data from LANDSAT, one of the original research satellites. At the same time other professional groups—doctors, journalists, and academics—have established computer networks with international access to national and local data bases.

The design of information technologies can encourage decentralization and flow of information, networks instead of hierarchies, process rather than product, co-operation instead of competition. Whether this potential becomes actual depends on non-technical decisions, which in turn will influence design and innovation (Figure 72).

## BIOTECHNOLOGY

The information technologies are only part of the 'high-tech revolution'. Recent developments in biotechnology and their rapid application has resulted, in a scant ten years, in a kind of explosion of knowledge and know-how. Genetic engineering and biotechnical businesses, sperm repository banks and fertilization clinics, have become part of our cultural landscape.

As a woman I find the technologizing of human reproduction disturbing. In-vitro fertilization, 'test-tube babies', surrogate mothers, 'sexing' babies (choosing the sex of the child), manipulating material in the embryo and fetus—this is a technology whose profound consequences in the lives of many people in the richer nations have not really been confronted. For instance, some women undergo in-vitro fertilization procedures, but only a few of them actually have babies. The low success rate of these experimental procedures does not match the high expectations promoted by biotechnicians and the media. (Renate Klein, a Swiss neurobiologist and co-author of *Test-Tube Women: What Future for Motherhood?*, estimated in 1984 that 85 to 90 out of 100 women leave the in-vitro fertilization clinic 'without child'.)

## Who spends most of the world's total research and development budget?

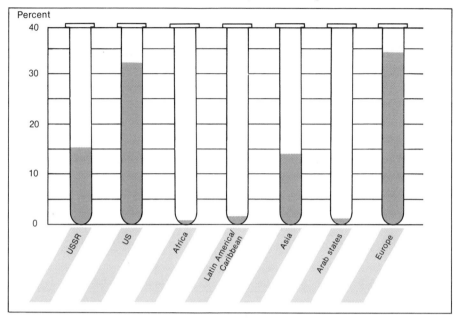

FIGURE 72. *Of the world's total research and development budget, 81 per cent is spent by the USSR, the US, and Europe.* (UNESCO Statistical Yearbook, 1987.)

And their ironic implications tend to be ignored: while a few babies are being 'created' at great expense in one part of the world, population density continues to be an urgent problem for poorer nations.

Experiments in the technologizing of human reproduction are accompanied by developments in modern agriculture, where genetic engineering is altering our ideas about species stability and integrity. In Nature, genetic material is rarely transferred from one species to another, but in the laboratory and then on the farm, transgenic transfer has resulted in artificially created creatures like the 'geep'— part-goat and part-sheep. While artificial insemination has been used for years in livestock production, the cloning of domestic animals is now a commercial reality. Embryos are cloned and some of them implanted in surrogate mothers for development. Others are frozen to see if they are worth developing later when the first batch has proven itself.

Living on a farm in Nova Scotia, I have been struck by how rapidly this technology is being developed and applied. Large industrialized farms have become, in effect, laboratories for testing new biotechnologies. The first US patent for a genetically engineered animal was granted on 12 April 1988 to a team of Harvard University geneticists who successfully introduced a cancer-causing gene into laboratory mice. Despite controversy, some nations permit testing of genetically engineered animal vaccines. In 1987 the US legally sanctioned the first environmental release of a genetically engineered bacterial preparation called Frostban, designed to protect a crop from frost damage. This was the first application of laboratory-created organisms.

## DESIGNER GENES

In plant-food production, geneticists are working to increase crop yields by imparting disease-, herbicide-, pest-, and stress-resistance, and by increasing crop quality and making crops easier to harvest. The following table shows the extent of this kind of enterprise.

### COMMERCIAL ACTIVITY TO INCREASE CROP YIELDS BY REGION

| TYPE OF PRODUCT | NUMBER OF COMPANIES | | | | | |
|---|---|---|---|---|---|---|
| | U.S. | Canada | Europe | Latin America | Japan | Total |
| Seeds | 137 | 14 | 38 | 3 | 11 | 203 |
|   Disease resistance | 40 | 4 | 15 | 2 | 8 | 69 |
|   Herbicide resistance | 26 | 3 | 8 | 0 | 1 | 38 |
|   Nitrogen Fixation | 20 | 1 | 6 | 1 | 0 | 28 |
|   Pest resistance | 18 | 2 | 4 | 0 | 0 | 24 |
|   Stress resistance | 15 | 3 | 4 | 0 | 1 | 23 |
|   Protein improvement | 18 | 1 | 1 | 0 | 1 | 21 |
| Plant Diagnostics | 54 | 3 | 19 | 4 | 1 | 81 |
| Plants Used as Foods & Feeds | 75 | 8 | 56 | 5 | 3 | 147 |
| Other Related Products | 10 | 2 | 12 | 25 | 1 | 50 |
| Grand Total | 276 | 27 | 125 | 37 | 16 | 481 |

(Also includes other advanced techniques)

SOURCE: *Bio/Technology*, Vol. 6, March 1988

'Designer genes' make new forms of life custom-tailored for specific sites and climates, and are already on the market, or are about to appear.

Frighteningly, much research reaches the applied stage before adequate testing can be done. The impetus? Competition within and between various scientific groups, and commercial competition on the 'free market' for genetically engineered products. And how do we

GENETICALLY MANIPULATED CROP
PLANTS AND PROBABLE YEAR OF
COMMERCIALIZATION

Rice . . . . . . . . . . . . . . . . . . . . . . 1991
Wheat . . . . . . . . . . . . . . . . . . . . 1992
Corn . . . . . . . . . . . . . . . . . . . . . . 1992
Soybeans . . . . . . . . . . . . . . . . . . 1992
Rapeseed . . . . . . . . . . . . . . . . . . 1991
Sunflower . . . . . . . . . . . . . . . . . 1991
Barley . . . . . . . . . . . . . . . . . . . . . 1992
Sorghum . . . . . . . . . . . . . . . . . . 1992
Alfalfa . . . . . . . . . . . . . . . . . . . . 1992
Fruit . . . . . . . . . . . . . . . . . . . . . . 1990
Tomatoes . . . . . . . . . . . . . . . . . . 1988
Potatoes . . . . . . . . . . . . . . . . . . . 1989
Other Vegetables . . . . . . . . . . . 1989
Sugar Cane . . . . . . . . . . . . . . . . 1989

SOURCE: *Bio/Technology*

*Different strains of rice developed at the International Rice Research Institute are planted on terraces in the Philippines. (World Bank Photo.)*

define 'adequate testing'? Are biotechnologists responsible for the short- and long-term effects of, say, the release into the environment of one of their products? Are policy-makers in a position to make informed regulations about new life-forms? And what about the general public? We shall have to live with the consequences, both good and bad, and yet many of us are content to 'leave it to the experts', even when we know these experts may have a conflict of interest between public good and private gain.

Optimists brush these questions aside, claiming that the current thrust of biotechnology could lead to the solving of the short-term food problems of an overcrowded world. But realists know that long-term global food security based on sustainable agriculture requires more than a technical fix—whose environmental effects are not known. And a recent statistic puts the problem of food shortage into perspective. In 1985 over a billion people in developing countries were starving or could not get enough to eat to lead active working lives. At the same time, intensive modern agricultural methods and technology, plus government subsidies in the richer nations, have led to food surpluses.

The methods and technologies of 'agribusiness' have caused soil erosion and degradation, pollution, and salinization of ground and surface water. In Canada, Senator Herbert Sparrow's 1984 report, *Soil*

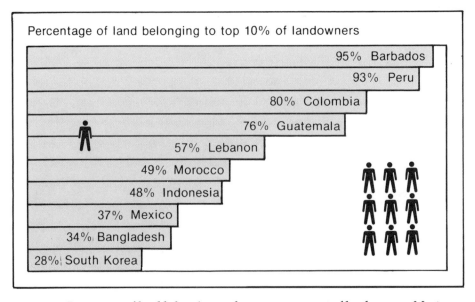

FIGURE 73. *Percentage of land belonging to the top ten per cent of landowners. Most rural households have practically no land.* (Book of International Lists, 1981.)

*Women produce sixty to eighty per cent of the food in Africa and Asia. Here they prepare the ground before planting rice in the Casamance region of Senegal. (World Bank Photo.)*

*at Risk*, suggests that capital-intensive farming and over-emphasis on productivity are robbing the soil of its fertility and the rural countryside of its people (see Chapter 9). Rural populations of the poorer but somewhat industrialized countries, like Mexico and Brazil, are migrating to cities, searching for jobs. Three-quarters of the agricultural households in many of these countries are either smallholders or landless. Ownership and control of land and land-use remains in the hands of a few (Figure 73).

In the past, international aid agencies have tried to ignore these political and social realities and have opted for technical fixes—for instance, building dams, or encouraging the use of chemical fertilizers and pesticides, without concern for side-effects—as the solution of the problems of Third World agriculture. They've also neglected human resources, preferring to think of 'the farmer' as male, although women produce sixty to eighty per cent of the food in Africa and Asia, and forty per cent in Latin America. This neglect has delayed or negated many of the agencies' attempts at agricultural reforms, as men are trained to do work that women have always done and continue to do—another indication that it is human, not technical, knowledge that is lacking.

*International-aid agencies tend to think of 'the farmer' as male. An extension agent demonstrates the use of a metal plough to men in the same region of Senegal. (World Bank Photo.)*

The global environmental consequences of these poor agricultural practices range from desertification and floods to potential world-wide climatic changes as tropical rain forests are destroyed. The loss of species diversity in the rain forest impoverishes the world's genetic resource pool, an irreplaceable survival resource.

## OUR COMMON FUTURE

Biotechnological advances cannot begin to touch the roots of the problems of global agriculture reform. This is why the 1987 Brundtland report, *Our Common Future*, suggests a switch of focus from technological to holistic. This approach would include the use of appropriate technologies—appropriate both in terms of scale and 'fit' with natural and social ecosystems (see Chapter 14). For example, small irrigation projects may be more useful and far less environmentally destructive than large dam projects in some areas. Small solar, biomass, or wind-energy projects may be more appropriate than nuclear reactors, which are capital-intensive, create high-level nuclear waste, and need a large output grid. A holistic approach would give priority to human needs and observe ecosystem protection.

Some cultural analysts, particularly physicist Fritjof Capra, also take a holistic and ecological approach, viewing cultural change in the light of general systems theory. Capra, a California teacher in both peace and conflict studies and of theoretical physics, focuses on the interaction of parts with the whole, and thinks in terms of dynamic process rather than structure. In describing the material universe, he uses the metaphor of a web of interrelated events, all in relatively stable patterns, in which no part is more significant than any other. The web of our relationship with each other, with technology and with the global ecosystem, would be non-hierarchical but have structured patterns. He emphasizes that what we see of the web depends on who is doing the looking. For instance, a technician, enchanted by the 'sweetness' of a technological solution, may forget about the complex realities of the whole field.

Nothing is static and absolute; everything is dynamic and relative, including technologies. This means we have a capacity, within the framework of overall systems, to make choices and changes. For instance, the shock of the 1970s' oil-supply shortages catalysed some interesting changes in the rich nations. First there was panic: gas-station line-ups, high costs, and possible home-heating oil shortages. This was followed by thoughtful reappraisal and deliberate change: energy conservation became a necessity and therefore a matter of government policy; homes and public buildings were insulated and smaller cars replaced 'gas guzzlers'. Existing technologies were adapted. New ones emerged. 'Soft-energy' alternatives began to look economically feasible—and even environmentally desirable. The sharp downward trend in energy/output ratios, for both industry and consumer services, throughout the industrialized world continued during the 1970s, contrary to earlier predictions of conventional energy experts. The fact that energy-use increased again in the late 1980s does not negate the fact that people can react creatively when they have to.

Now our use of fossil fuels is being questioned by the potential consequences of the greenhouse effect. As a result, nuclear power could become a dominant energy technology—with effects on the environment that could be as negative as those produced by fossil fuels. But it is clear that the whole matter of energy must undergo change, not just in technologies but in our values and priorities. A 1983 in-depth analysis, *2025: Soft Energy Futures for Canada*—published by the non-governmental organization Friends of the Earth—predicts that even our northern country, Canada—just by reducing waste—

could use less energy in the year 2025 than in the 1980s, at less cost to the consumer, and without significant life-style change. This would mean major government policy changes involving the decision to price energy at its replacement cost. Conventional fuels would become much more expensive if the real cost of production—including environmental degradation—was included in the price. Energy conservation would become more important, as would the use of renewable sources.

## CITIZEN SCIENTISTS

This kind of transformative change within a nation or nations cannot be imposed from the top. It requires consensus among informed citizens who involve themselves in planning the implementation of new technologies, as well as in formulating policy. The gap between scientific and technical experts and a lay public has to be bridged through mutual exchanges of information and experience.

It was the ecology movement that adopted philosopher René Dubois's phrase, 'Act locally while thinking globally.' We now have public hearings on the environmental impact of certain kinds of development, or in planning new hospitals. We could have public hearings to discuss and assess new technologies. At Cambridge, Mass., in the late 1970s the local community participated in the debate about whether laboratory experiments with recombinant DNA should be allowed in their 'backyard', and came to a negotiated settlement. This is just one example of how democratic processes can provide practical solutions.

The expanding role of Non-Government Organizations at the United Nations, and in nation states, suggests that achieving consensus and plurality may not be so hard. The new information technologies provide the means to reach millions and to transform, given the political will, global tensions into opportunities to create new structures to meet human needs. This is not a utopian fantasy. Technology has the potential for creating a world where everyone has enough to eat, enough shelter and clothing, joined with a sense of human dignity; where everyone has clean air and water, and the earth they walk upon is not poisoned or abused; where other life species are respected as co-inhabitants of the earth.

Technology has given us the means to realize this ideal situation. But consider these facts: approximately twenty-five per cent of the world's people consume seventy-five per cent of the world's wealth. The gap between rich and poor nations has been widening, nearly doubling between 1955 and 1980, with the thirty-six least-developed countries experiencing insignificant economic growth. As the Brundtland report demonstrates, sooner or later these inequities will bear upon us all. What happens to one strand of the web affects all the other strands. Decisions about technology, and technological decisions, are inseparable from the fields they are made in. Technical fixes may seem to be part of the answer, but they may also be part of the problem.

We return to our original image: the wind-surfer who is balanced between sea and sky; a creature of infinite adaptability, flexibility, and knowledge. Good surfers know how dark and powerful the waves are; know that a ship's wake could capsize and drown them. They also know that those who do not dance on the waves cannot be in touch with all the forces that impel and sustain them.

# Satellite Remote Sensing

The advent of earth observations from space has made possible the study of Earth as one system. The processes and phenomena that are now being studied—and will be studied in the future—typically occur over large areas, frequently over the entire globe. Furthermore, they are very dynamic, changing sometimes in minutes, sometimes in decades. Satellites are essential to study such processes.

The Earth surface was first observed with an artificial sensor in 1869, when Gaspard Félix Tournachon took pictures of Paris from a balloon. Aerial photographs became a valuable tool during both world wars, and techniques for obtaining information from them developed significantly before 1950. As airborne films and cameras, as well as interpretation techniques, improved in the 1930s, aerial photography began to play an important role in natural resource inventories such as soil and vegetation surveys, geological mapping, topographic mapping, and land-cover land-use mapping.

The modern discipline of remote sensing—that is, the science and technology for observing objects or phenomena from a distance—began in earnest with the launch of Earth-observation satellites in the 1960s. The first satellites were useful mostly for imaging clouds in weather forecasting. However, rapid improvements in satellite and sensor technologies led to the launch of the first Earth-resources satellite (LANDSAT 1) in 1972 by the US National Aeronautics and Space Administration (NASA). The multispectral scanner aboard this satellite could image the entire

*A composite image of vegetation conditions in North America in July 1988, produced with data from an Advanced Very High Resolution Radiometer on board the National Oceanic and Atmospheric Administration satellite. Images from the last ten days of July were registered with a spatial resolution of one kilometre; the most cloud-free images were selected, and a new image with the least amount of clouds was created. The photosynthetic activity of the plant canopy was then emphasized, with lighter tones indicating greener vegetation. Changes from year-to-year and seasonally can be monitored this way. (Courtesy the Canada Centre for Remote Sensing: Surveys, Mapping and Remote Sensing Sector, Energy, Mines and Resources, Canada.)*

globe every eighteen days, at a high resolution (each picture element covered an area about 0.4 hectare in size), and in various spectral bands. LANDSAT 6 is now under construction. In 1986 the French Système Probatoire d'Observation de la Terre (SPOT) was launched, the first in a series of high-resolution satellites (picture elements 0.01 hectare in size). The next year the Japanese Marine Observation Satellite (MOS-1) was launched. By 1989 data from all these satellites were received by ground stations around the globe.

The techniques of remote sensing are based on the fact that various types of objects and materials respond differently to incident electromagnetic radiation. For example, when solar radiation reaches the vegetation canopy, some of it is reflected into space, some is absorbed by the canopy during photosynthesis, some is transmitted to the ground, and some is emitted back into space. The magnitudes of these fractions vary with wavelength depending on the object or material. By judiciously choosing the wavelength bands where measurements are taken, a great deal of information may be obtained about the object.

When one wishes to observe the earth's surface, the spectral bands must be located in regions of the electromagnetic spectrum where the atmosphere is transparent. In the optical part of the spectrum, several such windows are available, including the blue-green-red portion where the human eye is sensitive. When the atmosphere itself is the object of interest, the bands are placed where the atmosphere is opaque to different degrees. This yields the ability to 'probe' the property of interest—for instance, the atmospheric temperature profile.

Imaging radars and other microwave sensors can operate successfully under cloudy conditions. In addition, since radars provide their own source of illumination, they can observe the Earth's surface at night or during the winter in polar regions; this feature is very important for scientific studies as well as for practical resource-management applications.

The digital data recorded aboard the satellite are transmitted to a ground receiving station at very high volumes. Each station is capable of receiving data from a large area of the Earth's surface. For example, a station in central Saskatchewan—operated by the Canada Centre for Remote Sensing, Department of Energy, Mines and Resources—covers most of the North American continent. The data are recorded at the station on high-density digital tapes, and corrected for distortions due to spacecraft motion, the shape of the Earth, sensor calibration, etc. The corrected data are then made available to users.

While various clues—such as tone, colour, texture, shadows, size, shape, and height—are available to a photo interpreter, 'colour' assumes key importance in satellite remote sensing. Radiation entering the lens of a multispectral sensor is decomposed into separate spectral channels (bands), and its amount is recorded digitally for each band. The multispectral scanner on LANDSAT 1 had four spectral channels (green, red, and two infrared), while current experimental state-of-the-art airborne sensors have over 200 spectral channels.

Since the satellite data are digital, computer-based analysis techniques are most appropriate, and are supported by the rapid growth in personal computers and desk-top workstations. These tools magnify the ability to analyse data, and to merge data from various sensors and from other sources (airborne, surface, historical, statistical). Finally, we can use the data as inputs into models that aim to describe, explain, and ultimately predict the processes or phenomena under consideration. This area, including the merging of remote-sensing and geographic information-system methodologies, is developing very rapidly.

The study of global change brings high demands on remote sensing in the areas of sensor development and data acquisition; data preprocessing, storing, and retrieval; data analysis and integration; and input into process models. A key NASA program, the international Earth Observing System (Eos), will in the late 1990s launch several satellites with many sensors to probe, over a number of years, environmental characteristics ranging from the Sun to the interior of the Earth. The raw data alone provided by these sensors could fill 7,500 computer tapes a day, three million per year, according to initial estimates. Advanced storage media will make it possible to keep data from the previous two years on line. The principal aim of Eos is to contribute to the study of the Earth as one system. Individual countries are also planning new satellites that will provide data for global change studies—including the Canadian RADARSAT, with its advanced imaging radar, due to be launched in 1994. The implementation of systems to process, archive, and use data from these satellites will be a major challenge for the next two decades.

JOSEF CIHLAR

# Reflections on Science and the Citizen

One of the joys of my professional life has been my work with citizen groups on issues related to pollution, energy conservation, the arms race, or research priorities. Freed from the competitive shackles of the compulsory educational system, my neighbours—mostly women—learned easily and quickly. They were eager, helpful, and co-operative, always ready to share knowledge and resources. I had no problems in explaining quite complex chemical and physical phenomena. These were retained and used not only in terms of facts that entail certain consequences, but also in terms of the underlying chain of reasoning and evidence. A number of these citizens have become able interveners at public hearings and in themselves are now resources for others.

Based on my own experiences, I can say that the key elements in the acquisition of scientific knowledge by citizens—knowledge that makes them responsible citizen scientists—are strong motivation, confidence in their own common sense, and a non-competitive atmosphere that ensures that all participants are both teachers and learners. This is not difficult to achieve.

As an academic I have learned in this process of working with citizens' groups that the link between doing science and understanding science is much more tenuous than I had previously assumed. It is possible to understand, in considerable depth, science and its impact without actually doing science. Conversely, doing science—i.e., conducting successful research—does not necessarily imply that researchers have an understanding of what they are doing in a broader context.

Two facets of what we call science may be worth reflecting upon here: one relates to the scientific method, the other to the notion of a scientific field or discipline and the work of its practitioners.

Francis Bacon's science, new in the seventeenth century—with its experiments, its requirements of internal consistency, and its aim of discovering universally applicable laws of nature—carried within itself the separation of knowledge from experience. Extracting the general from the particular allowed knowledge to be accumulated and spread about the world faster. Thus the last two or three hundred years have seen an unprecedented growth of knowledge and information. However, the shadow side of this historical development cannot be overlooked. First of all, any general law intrinsically lacks a context; this limits the law's usefulness as a sole guide to specific applications. Then there is the emphasis on abstract knowledge over concrete experience. This has dramatically lessened the confidence of people in the astuteness of their own senses. There are many today who prefer not to rely on their own direct experience, but look for 'experts' to tell them what it is they sense, feel, or hear.

Secondly, in terms of the facets of science, one has to raise the question of what defines a scientific discipline and its practitioners. The answer to 'What is physics?' has always been 'Physics is what physicists do.' And the content of physics, as seen in teaching and research, has changed greatly over the years. Much of what constituted physics at one time is now regarded as a domain of mechanical or electrical engineering or as applied science. But the discipline is still defined by the problem areas addressed by its practitioners, however restricted or extended they may be.

Defining a scientific discipline by the activities of its specialists gives such scientists a gatekeeper role. *They* will admit or reject both people and problems in an 'us' versus 'them', or an 'inside' versus 'outside', mode. The exclusion verdict of 'This is an interesting approach, but it really is not science/physics/biology . . .' is well known to all those who have tried to bring outside knowledge to inside reasoning.

The participation of concerned citizens in political decisions with significant technological or scientific components has been greatly held back by these two facets of science. Often citizens bring direct experience to the discussion with experts, only to find that this experience is undervalued or discarded entirely in favour of non-contextual and abstract information. Usually citizens are outsiders *vis à vis* specialists and experts, who tend to equate a particular career path with the possession of relevant knowledge.

In my opinion these hurdles are perceptual rather than real, but the practical impediments to citizens' participation are very real indeed—although there are relatively simple ways to overcome them, including the work I described with citizen groups.

The task of the future is to build knowledge and understanding among and between citizens and scientists, so that the distinction between the two groups vanishes—so that *both* become citizen scientists, potentially able to solve our problems together.

URSULA M. FRANKLIN

CHAPTER 13

# The Missing Tools

## ROBERT GOODLAND and HERMAN DALY

The central preoccupation of economics today is growth—by which is meant an expanding scale of resource uses. Growth is offered as a cure for most economic and social ills. The implicit aim of economic growth, as pursued all over the world, is to produce and consume on a per capita level equal to that of the US or Western Europe. But is this realistic? It takes about one-third of the current annual world extraction of non-renewable resources to support at its current level the five per cent of the world's population living in the US. This means that the flow of current resources could at most support only fifteen per cent of the world's population at the US consumption standard, with nothing left over for the other eighty-five per cent!

If the entire world were to consume resources at the US rate, the total flow of resources would have to increase roughly seven times. Considering the damage that even our present rate of resource use wreaks on our global environment, it seems unlikely that we could get away with a sevenfold increase. How did this delusion come upon us? What view of the world and of economics has made the present absurd and unjust economic arrangements possible?

Adam Smith gathered a lifetime of observations of the European economy into his famous book, *The Wealth of Nations* (1776). In his vision, as elaborated in today's textbooks, the economic process is a circular flow of money and goods between families and firms in an isolated system. Families spend money to buy goods that firms produce, and firms spend money to buy factors of production (land, labour, and capital) owned by households. As in Figure 74, money flows clockwise in payment for real things (goods and factors).

The focus in this vision is on two decision-making units: firms and families. The theory explains many things: how supply and demand works, which explains the setting of prices, which in turn explains the allocation of resources and distribution of income. In national econo-

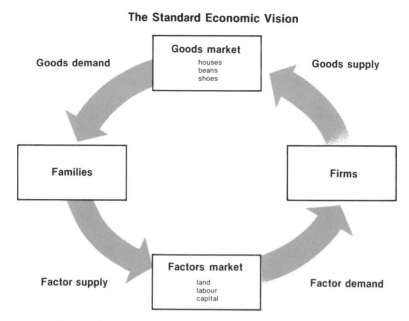

FIGURE 74. *In the standard economic vision, money flows clockwise in payment for real things.*

mies the total volume of the circular flow is described as the national product. Maintaining growth of the gross national product, or GNP, has become the primary focus of economic policy. Many international publications—including those of the World Bank, for example—rank countries according to their per capita GNP. We tend to assume that when GNP grows, human welfare will also increase.

This vision has guided economists in their understanding of the market economy ever since Smith's day. It has been valuable, as it explains how the industrialized economies today can produce efficiently the enormous variety of goods, in quantity, that most citizens of the developed world now take for granted. But it has not been able to take proper account of many aspects of economic activity that we see today. Consider again Figure 74. The circular flows of product and production factors are isolated from the environment. There is no way, in this construct, for economic activity to be affected by environmental changes. For Smith this did not make much difference. In his world the human economy was tiny; a factory employing ten men was noteworthy. For economic theorists the environment was simply an infinite source of raw materials, and an infinite sink for wastes, because it was so large compared to the economy.

But this assumption should have been replaced decades ago, when the increasing number of human impacts on the global environment was first felt. When we look today at the kinds of impact we have on

*The tropical forest is dwindling at the rate of 30,000 hectares a day. Trucks are loaded with lumber in Gabon, West Africa. (World Bank Photo.)*

the environment, we notice immediately that they are not all the same. Some are reversible; others are not and cannot be cured by throwing money at them, or even by political will.

The best-known examples of reversible impacts are cases of industrial pollution that have been cleaned up. Ohio's Cuyahoga River was treated as a petrochemical sink and died decades ago. It was so polluted that it occasionally even caught fire. But after the political will and financing were mustered, the pollution was reversed, and the river today is attractive and clean. The irony is that it may be far more expensive to cure environmental damage than to prevent it in the first place. The discipline of economics has contributed greatly to the mechanisms by which such impacts can be cleaned up—as in, for example, the 'polluter pays' principle, and in incentives, permits, taxes, rights trading, and penalties.

Many of our impacts on the environment, however, are irreversible, or are only remedied over a very long time. They include climate change, ozone depletion, contamination of groundwater, soil erosion, tropical deforestation, and species extinction. Once the stratospheric ozone is depleted, it cannot be restored except possibly over the very long term; groundwater, once contaminated, can never be cleansed; species that become extinct can never be restored. The damage may be permanent.

Consider two impacts: tropical deforestation and species extinction. The tropical forest is dwindling at the rate of 74,000 acres a day. This habitat supports half the world's five to thirty million species on seven per cent of our land area. Latin America alone may support fifteen million species, possibly one third of the world's total. Jag Maini has discussed the causes of forest loss in Chapter 8: (i) the clearing of vast areas for cattle ranches or speculation—thousands of square miles of tropical forests are sacrificed to give us a few years of minutely cheaper beef hamburgers; (ii) demand by the industrialized world for tropical forest products; and (iii) poverty (poor people forced into shifting and temporary cultivation of cleared tropical forest).

European forests tolerated decades of pollution with little signs of stress, only to succumb over vast areas within a few years, beginning around 1980. This suggests one other observation to keep in mind: that the environment often appears able to absorb long-term abuse— then suddenly, with little warning, succumbs or overshoots on a global scale. If we ignore the capacity of the Earth to absorb the demands we place on it, we may not see the consequences immediately. But once they do appear, they may, as with European forests, be catastrophic.

If economic growth continues, its impact on the environment inevitably increases. The growing stream of products demands a growing

*Industrialized countries produce millions of tons of trash every year, some of it toxic, most of which is buried. (SSC/Photo Centre/ASC.)*

quantity of materials and energy, and these flows through the economic system now rival some natural cycles. Inputs from human activity produce many atmospheric contaminants—such as soot and carbon dioxide, sulphur compounds in the northern hemisphere, and several trace metals—and are now larger than natural flows. Moreover, our sink is getting full. The United States produces 150 million tons of trash each year, and ninety per cent of this ends up buried. In 1987, 3,186 tons of New York garbage were barged 6,000 miles for four months, refused by six states and three nations. Finally the load was incinerated, but even the resulting ash may be toxic.

Economic activity is now understood to have enormous impacts on the environment—some that we can hope to repair, and others that are irreparable; some that we can see now, others that will become apparent only to the next generation. As a discipline, however, economics is ill-equipped to understand the ties between economic activity and environment. Strange though it may seem, economists and ecologists rarely meet. When they do, they have difficulty communicating because they speak different languages; one might say they inhabit different worlds. In most introductory economic textbooks, the environment is relegated to a discussion of externalities (the effects of the outputs of goods and services on those who are not buying or selling or using the goods), and externalities are seen only as a minor imperfection in the circular flow of production and consumption.

## TOUGH QUESTIONS

We must accept that not everyone can consume at the level of the wealthy Western countries. We must either admit that economic development is just for a lucky few, or redefine development so that it can apply to everyone living now as well as in the future. This means a *convergence* towards a sufficient level of consumption in all countries. An adequate standard of living in poor countries requires scaling down in the wealthy. But where does the equitable point occur? This is a question that standard economics does not begin to answer. The market measures the relative scarcity of individual resources, and through the price system, can optimally allocate these resources among alternative uses. But just as there is nothing in this system that can determine an ecologically sustainable scale of resource use, so is there nothing to identify the most equitable distribution of resources.

Our reliance on GNP as a major tool of standard economics, as a primary yardstick of progress, is an obstacle. This is because GNP measures not the satisfaction of wants or needs but simply production—for any purpose, resulting from any activity. Depletion of mineral reserves and ecosystems is counted on the same basis as sustainable use of resources. A nation encouraged to maximize GNP may choose to 'liquidate' its natural-resource base by rapidly cutting down forests, or mining its mineral resources, and will get brownie-points for doing that from many international development agencies. Moreover, rapid obsolescence of consumer products increases the GNP. If A buys five cars, each lasting only two years, and B buys one car that lasts ten years, both get ten years of transportation, but A contributes more to GNP.

Reliance on GNP makes it difficult to measure the costs of economic growth, which, beyond some point, brings with it disruption of ecosystems, and hence a reduction in their services. These services are often part of our global commons, and include good climate, soils, air, and water, and genetic diversity. Some can be replaced, although at great cost. Declining soil fertility is masked by the use of more machines, energy, chemical fertilizers, and pesticides. But the costs of such actions—protecting buildings from acid rain corrosion, for instance, or maintaining shorelines eroding with rising sea levels due to greenhouse warming—simply boost GNP. Conventional measures of progress, such as GNP, provide no means of comparing the marginal costs and the real benefits of growth itself. Instead the cost of repairing ecosystem damage is counted as an economic benefit, while irreversible effects may be completely ignored. Eventually growth costs more than it is worth.

## DISCOUNT RATES

Another impediment to environmentally sound development is high discount rates. (The discount rate is the opportunity cost of capital, or the investment rate of interest, which is set by the market but influenced by government policy.) For most people, future costs and benefits are of less concern than present ones; we decide about tonight's dinner before we worry about next week's menu (Figure 75). In economics, a high discount rate gives greater weight to today's concerns, while a low discount rate places more stress on tomorrow's. High discount rates, often used in planning investment, encourage rapid exploitation of resources and ecosystems for immediate payoffs, discourage focus on long-term benefits, and minimize future costs and risks. If a whaler, for example, can make a profit of fifteen per cent per year by exterminating whales over ten years and then investing the proceeds in another activity, what economic incentive is there to make a profit of only ten per cent per year by harvesting the whales sustainably?

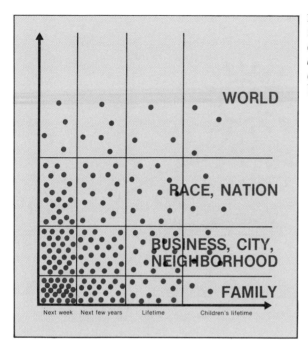

FIGURE 75. *Human perspectives. Most people are less concerned about future costs and benefits than present ones. (Dennis L. Meadows, Limits to Growth, 1972.)*

WORLD

RACE, NATION

BUSINESS, CITY, NEIGHBORHOOD

FAMILY

Next week    Next few years    Lifetime    Children's lifetime

Rates that support sustainability could be set at zero for some activities. Then the profit would be equal to the annual, or sustainable, harvest, less the cost of harvesting. If there were *no* profit to be made in harvesting whales, then no one would do it.

The economics that guide today's decision-making of governments and business does not view the economy within a finite environment. The contrast between our personal behaviour and the behaviour of our society is striking: as individuals many of us already drive smaller cars, conserve energy, recycle waste, provide for our descendants, and behave thriftily. As a society, though, we allow ourselves and our globe to be operated on a system of frontier economics created when the world was essentially empty of people; when resources were scarcely tapped, air and water were clean, and land and forests were abundant. This idyllic era has long since vanished, but orthodox economics hasn't kept up; it needs to be modernized to reflect today's full, or overfull, world. A revamped economics will accommodate environmental and ethical concerns, and provide a basis for sustainable development, aimed at achieving a reasonable and equitably distributed level of economic well-being that can be maintained for many generations. The old theory of growth was politically convenient. It allowed us to avoid two essential but politically sensitive

# The Steady-State Economic Vision

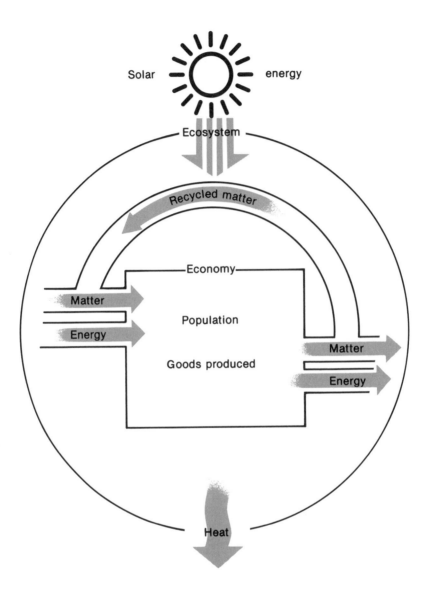

FIGURE 76. *In the new economic view, the environment is finite and the economy is a subsystem of the environment.*

issues: redistribution—sharing the wealth—and population control. Both are alternatives to growth as a solution to poverty, but both are politically touchy.

What would sustainable and ecologically sound economics require? First, to acknowledge that the economic system is only part of a much larger system that includes our global environment. The view of our economy as an isolated circular flow of products and values needs to be replaced by an understanding of two unalterable facts: production and consumption are based on the taking of matter and energy as raw materials from the environment and processing them; and this eventually leads to releasing waste and pollution into the environment (Figure 76). Since the environment is finite and the economy is a subsystem of the environment, it is obvious that physical growth in the economy cannot continue forever.

Anyone who asserts that there are limits to the activities of our economy is soon presented with a litany that someone once said could never be done but subsequently were done. Certainly it is dangerous to specify limits to knowledge. But it is equally dangerous to assume that new knowledge will abolish old limits faster than it discovers new ones. The discovery of uranium expanded our resource base, but the subsequent discovery of the dangers of radioactivity contracted it. So, too, is it risky to assume, as many economists continue to do, that when a natural resource becomes scarce, we will always be able to find a substitute. Sometimes this is possible, as in the substitution of aluminum wires for copper, to be replaced in turn by fibre optics. But more often substitution of natural capital by man-made capital is only at great environmental cost. Floods and drought related to deforestation of a watershed can be remedied by building dams and dikes, but this is rarely as effective as the original forest and can cost far more than was earned by deforestation. Similarly the replacement of natural soil fertility by artificial fertilizers, while technically possible, is often more expensive, and can result in runoff water pollution, risks to the ozone layer, and contribute to greenhouse warming.

If growth cannot continue forever, then it will have to stop somewhere. At some point the economy will have to attain a steady state, supported by constant stocks of things and people. In a steady state these stocks are not static: people die and are born; things wear out and are replaced. But in a steady state there is no net growth in population, and no net growth in the extraction of matter and energy from the environment. Knowledge and technology, however, are not held constant. Neither is distribution of income nor the allocation for resources. The steady-state economy is not zero GNP growth. It can develop in quality but not grow in quantity, just as the Earth, of which the economy is a subsystem, develops without growing. The assumptions of standard economics notwithstanding, the surface of the Earth does not grow at a rate equal to the rate of interest.

Given a steady-state economy, the question is not one that can be settled through the market. It transcends economic calculations and is an ethical and social issue—a matter on which a collective, democratic decision must be made. In deciding on the size of our economy, we are making two value judgements: how far to base our economic system on the 'takeover' of habitats of other species; and how much to 'draw down' geological capital, which, if used now, is not available to the future. Obviously, our present demands on the environment cannot exceed some level, but neither should they be reduced below some level. Our present generation has legitimate needs, many of which are not being met for most people. We must strike a balance between the needs of the present and those of future generations.

Although we cannot predict future needs, we can prudently assume that future needs for natural resources will not be markedly less than our own. The sustainable use of natural resources therefore implies:

—Using renewable resources in ways that do not degrade them (harvesting renewable resources—whether whales, forests, or coral reefs—on a sustained-yield basis rather than 'mining' them to near-extinction).

—Using mineral reserves in ways that give easy access later. (It will be easier to use today's scrap metal tomorrow if it is recycled rather than dumped indiscriminately.)

—Introducing an orderly transition to renewable energy sources (including solar, biomass, wind, and hydroelectric) as non-renewable energy resources become too expensive to maintain with their slow rate of depletion.

—Maintaining natural soil productivity in agriculture (replacing what is taken out) while counting in the 'external' costs (greenhouse warming, reduced well-being) generated by the use of imported inputs like diesel, biocides, and fertilizer.

—Maintaining the resilience of agricultural systems in the face of changes, whether economic, climate, or species diversity.

## THE NEW TOOLS

The options of future generations cannot be left solely to economic mechanisms. Economics that includes the environment, however, can contribute tools to help identify the path to sustainability. Importantly, it can provide a replacement for GNP as a measure of economic progress.

Instead of keeping one account, the GNP, we should keep three:

(i) a 'Benefit Account' to measure the value of services provided by the economy and the environment;

(ii) a 'Cost Account' to measure such quantities as the degree and cost of depletion and pollution;

(iii) a 'Capital Account'—an inventory of the accumulation and distribution of stocks of both producer and consumer goods, including natural capital such as mines, wells, and ecosystems.

A major benefit of measuring the status of both the economic system and the environment is that we will be enabled to distinguish between activities supported by unsustainable exploitation on the one hand, and sustainable activities on the other. These accounts together can indicate environmentally sound levels of resource and energy use. With separate accounts for costs and benefits, we could balance the extra benefits of further consumption against the extra costs. This is a long-term, Utopian solution. In the short run, perhaps the best we can do is correct existing national accounts by subtracting depletion of natural capital from national income. For instance, deforestation, now counted as income, would be shown as what it is—a reduction in a country's wealth.

## THE WORLD BANK

The World Bank lends money to developing nations, using funds raised by floating bonds in wealthier countries. One of the most influential actors in international development, in 1988 alone it loaned $19.2 billion US for projects all over the developing world. Although it is set up for mega-projects, changes in the policies and practices of the World Bank show us how economic and environmental concerns can be integrated internationally. Since 1987 the Bank has become more capable of assessing the environmental aspects of development. A new Environment Department was created, with environmental concerns at the centre of policy-making, and an environment division was set up in each of the four geographic regions.

These developments reflect a new awareness at the Bank of the ties between economics and environmental concerns. As the president of the Bank, Barber Conable, wrote in 1986, '. . . the goals of poverty alleviation and environmental protection are not only consistent, they are interdependent. Sound ecology is good economics.' This new awareness is a significant change from the recent past, when Bank projects sometimes had bad environmental consequences. It is being expressed in concern with rapid population growth and the concept of sustainability.

Integrating environmental concerns into specific development projects from their earliest stages will eliminate most environmental risks and minimize remaining impacts, or even enhance environmental benefits. Unavoidable impacts are compensated. For example, if a reservoir cannot avoid flooding a habitat, then a similar tract will be purchased and conserved elsewhere.

New environmental policies cover pollution control, industrial safety, biocides, and reservoirs. Another policy, 'Wildlands', helps preserve endangered species and their habitat. So far the Bank has financed conservation in forty projects in twenty-six countries, protecting more than 60,000 square kilometres. The biggest is in Amazonian Brazil, where the Bank is financing the protection of 19,000 square kilometres of tropical forest in Rondonia province.

The Bank is also promoting environmentally sound practices beyond its specific financial influence. In 1986, for example, it lent for improved environmental procedures the entire Brazilian power sector. The borrower, Eletrobras, now has an in-house environmental department, an independent environmental review panel, and a master-plan to maintain environmental quality throughout its operations. The Bank is also working on improving the environmental expertise of borrowing governments, and on fostering participation in development projects by non-governmental organizations.

The Bank routinely discusses the economy with each of its member countries throughout the year, now emphasizing the long-term quality and sustainability of development. 'Many direct governmental subsidies are unsound both in environmental and in economic terms,' says Shahid Husain, Vice-President of the World Bank. 'They add to a country's fiscal burden, encourage the wasteful use of scarce resources and frequently benefit the large landowners.' New forms of national accounts, like those we have discussed, are being devised to measure depletion of non-renewable resources and unsustainable exploitation of renewable resources. The environment will be integrated into the economic dialogue that the Bank has with each borrower.

As with any large organization, changes in World Bank policy and practices do not occur overnight. The actual behaviour of large institutions, whether banks or governments, often lags behind changes in policies, and in the personal behaviour of concerned individuals. But the Bank is demonstrating how understanding the ties between economics and the environment must be integrated into economic practices.

## INTER PARES

At the other end of the scale from the World Bank, with its huge loans, is the whole grass-roots development movement, offering direct aid to people, often on a village level. Individually, these non-governmental agencies have next to no financial clout, but they have the advantage of flexibility and close contact with their developing-world projects.

Inter Pares (between equals), a non-profit charitable organization based in Ottawa, has supported community-based development overseas and in Canada since 1976. The loans made by the World Bank in 1988 amounted to 19.2 billion dollars US, the equivalent of about one week of the US defence budget. The project expenses of Inter Pares in 1988 were 2.2 million dollars US, about one minute of the military budget. Though so different in size, the avowed principles of the two development agencies are nevertheless similar. In supporting development groups and initiatives, Inter Pares requires participation of 'partners' (the people who are to benefit), involvement of women in the development of their communities, respect for cultural values, and sustainability. Development programs must lead to self-sufficiency and be ecologically, organizationally, and financially sound.

'There has been a tendency to blame the deterioration of the environment on short-sighted national development policies,' says the 1988 Annual Report of Inter Pares. 'What we often fail to recognize is that environmental destruction . . . is in part a result of development and economic policies which are vigorously supported by First World interest. The debt crisis, for example, has forced many nations to carelessly exploit resources.'

Inter Pares is one of a growing number of organizations calling for sustainable development based on good ecology. Its Third World work aims to demonstrate alternatives to environmental exploitation. Partners in Africa and Central America work to preserve traditional seed varieties. In Africa, Asia, and Central America, partners develop organic agricultural practices. Groups in Malaysia monitor the use and abuse of dangerous pesticides. In India Inter Pares supports reforestation and land reclamation; in Africa the development of renewable energy technologies; in Canada techniques for assessing the environmental impact of projects.

We need all these economic perspectives and tools to realize the environmental consequences of our actions, to re-focus economic behaviour away from short-term gain and infinite growth, and to preserve our own prospects—those of later generations and of the ecosystems upon which we depend. At best, unsustainable development is a huge gamble that is banking on the hope that by the time known resources run out, alternatives will be available. There must be an alternative to running the Earth as if it were a business in liquidation.

# The Hamburger Connection

If cattle gain 50 kg per hectare per year, and are slaughtered after eight years, and half the weight is non-meat (skin, bones, etc.), then each cow produces 200 kg, or 1,600 hamburgers.

It takes one hectare of cleared tropical moist forest, turned to pasture, to feed that one cow and to produce the 1,600 hamburgers. Because the land is fertile for grazing for only a few years, this is a one-shot deal, and an expensive one. After ten years—a generous estimate of the life of the soil—the return on the land from the hamburgers it produces will have been $3 US per hectare per year.

The cumulative effect of this hamburger consumption is equivalent to millions of years of evolution, and to thousands of species.

Conversion of all of Amazonia—4 million square kilometres—to cattle pasture would produce one month of hamburger for the world's population and *no more*—since the soil would be depleted and the forest irreversibly lost.

Adapted from J.O. Browder:
*The Social Costs of Rain Forest Destruction:*
*The Hamburger Debate.* Interciencia, Caracas, 1988

*It takes one hectare of cleared tropical moist forest in Brazil, turned to pasture, to feed one cow and produce 1,600 hamburgers. (World Bank Photo.)*

# Industry's Response

I do not believe that we are looking at an 'either/or' situation in terms of regulations and incentives to protect the environment. The problems the world faces in changing the policies, technologies, and lifestyles we have developed—particularly over the last two centuries since the industrial revolution got underway—are enormously complex. The solutions to the problems that confront us now will be equally complex, and it would be naïve to think there is one simple answer that will provide the necessary motivation for change.

Industry is going to be obliged to play an important role in implementing solutions, which in part will be driven by regulations. But solutions will also grow out of industry's responses to irresistible forces that will be exerting an ever-stronger influence on society: these will include pressures arising from greater understanding of ecological and health impacts, consumer wishes, changing market demands and changes in technology to satisfy those demands; growing awareness of societal needs and ecological limits; internally and externally generated pressures to behave as good citizens; and the need for the developed world to set examples for, and to aid, developing world projects.

It is important that business meets these challenges positively, without affecting in a major way the need to remain competitive and to continue to demonstrate good bottom-line performance. Here government incentives can enhance industry's natural healthy self-interest and provide additional motivation for change. Once embarked on this course, industry can make changes faster and more profoundly than any regulatory body.

Without doubt there is a place for regulation. One need only look at the ecological disaster of the Exxon Valdez oil spill off Alaska to see the need. That is a horror story, and there are many others in the same category. In that context I fully support the blunt instrument of regulations, which will ensure that proper care and attention are taken by all of us—not only the business collectivity, but individuals—to preserve our all-too-vulnerable ecology. But until recently, regulation and legislation have been essentially the only game in town. Now governments are looking for new tools to achieve more efficient and effective environment/economy integration: economic mechanisms such as contaminant charge schemes; tradeable emission discharge rights;

investment tax credits for exceeding environmental standards; reduced-interest bonds; economic incentives to promote effective environmental protection; systems for measuring the contribution of the environment to the national economy and national wealth; more research; specific projects demonstrating environment/economy integration. It is essential that industry participate, and even take leadership, in all these challenges—going beyond the minimum standards that have been established by regulations.

The concept of incentives to industry to make changes may be distasteful to some. It runs counter to the 'Polluter Pays' slogan. But that is an old-fashioned idea. It fails to recognize that we are all polluters—whether by putting gas into the automobile tank, turning up the thermostat on a cold day, having a beer out of a can (even a recyclable one), or in a hundred other ways. We all contribute to the pollution load and we are all going to have to pay, one way or another—in the near future through increased taxes, and over the long haul through being forced to adjust to new processes, new habits, and changes in our way of life.

ROY AITKEN

# CHAPTER 14

# Surprise and Opportunity:
# In Evolution, in Ecosystems,
# in Society

## C.S. HOLLING and STEPHEN BOCKING

The Chinese ideogram for change—for 'surprise'—combines two characters. One is crisis and the other is opportunity. The Chinese concept of surprise provides opportunity: a time for profound evolutionary change—or devolutionary collapse. We would like to look at the evolution of life on earth with this ancient wisdom in mind.

The world, and life in the world, have been generated and transformed by surprise from the beginning. (We are using the word 'surprise' to mean change that could not, on looking back, have been anticipated.) But now the surprises are changing in kind and intensity. What we are living with is a profound acceleration of the rates of change. Climatic change, acid rain, holes in the ozone, all are signs of this, and they are also signs that humanity has become a planetary force.

Earlier chapters have told us that the earth is a heat engine, constantly in motion and constantly changing. It has experienced volcanic episodes, drifting continents, changing oceanic currents, and ice ages. All these changes—some slow, others rapid—have influenced the course of evolution.

In addition the earth has been periodically bombarded from space by large asteroids that hit with a size and energy that caused global paroxysms. Many geologists believe that about 65 million years ago much of life, including the dinosaurs, was extinguished as the planet recoiled from an impact by an asteroid perhaps 10 kilometres in diameter. We are still experiencing the consequences. They included a global veil of dust and moisture that for perhaps ten years changed the temperature characteristics of the world and darkened it enough to prevent photosynthesis. Some eighty per cent of all species then on earth vanished, some very suddenly, others withering away slowly. For those that survived, the impact opened enormous opportunities for new plans of organization, new combinations. This is what fundamental evolutionary change is about: the creation of new environments that catalyse profound adaptive change.

Surprises can also arise from inside a system. Life and life's products, from bacteria to human institutions, both initiate and respond to change. But they also defend the status quo, sometimes to the point of stagnation, or to a fragile brittleness. Then, inexorably, a break occurs, and a surprise is generated within the system.

Another phenomenon of change has emerged relatively recently. That is the shift in our atmosphere as carbon dioxide and other trace gases, such as methane, increase at one to two per cent a year. In this case the agent of change is human industry. After only two centuries of industrialization the world environment is being transformed. As in the past, this surprise is having profound global consequences. We would like to explore those consequences with the question of sustainable development in mind, acknowledging the inevitability of change and considering if there are other aspects that must be maintained in order to ensure continued opportunity. If so, what are they?

Why should we do this? The appropriate units of life—for questions of sustainable development—are ecosystems: interacting assemblages of plants and animals, transforming and being transformed by the physical and chemical environment that envelops them and of which they are a part. On a global scale we see profound evolutionary change, true bursts of structural innovation. When we narrow time and space to the scale of ecosystems, humankind begins to take a role as one of the actors in the play. As in any good play there is a rhythm of slow change, development of crises, release, resolution, new stages of innovation. The play unfolds not in millennia, as in the global systems, but over centuries and even decades. We identify four acts to this *play*: initiation and pioneering, conservation, collapse as a result of some surprise event, and adaptation to the surprise. The process is repeated, again and again (Figure 77).

Take a forest ecosystem. During the initial pioneering phase, as when a newly abandoned field begins to return to the forest, opportunistic colonizing plants rapidly establish themselves. In doing so, they slow or reverse loss of soil nutrients, prevent erosion, and moderate the microclimate. That is followed by a phase of consolidation. Nutrients and biomass are slowly accumulated by the photosynthetic activity of plants, and the original pioneering species are slowly replaced by other larger, more competitive trees.

As in the evolutionary stage, both internal and external change can take place. Forest fires, for example, often seem to be external surprises, as do hurricanes, wind damage, disease, insect-pest outbreaks. But the more we examine the mature forest or ecosystem, the more we realize that increasing linkages of competition and predation make them so stable, so dependent on conditions remaining as they are, that they become brittle. They invite accidents or surprise. If the accident is not fire, it could be insect infestation—or pest outbreak or disease. The eternal paradox of order emerging out of chaos is matched at this phase by chaos as the inevitable consequence of order.

Those two phases of pioneering and consolidation are slow. When surprise happens, some resources may be destroyed, but much accumulated capital is suddenly released for other kinds of opportunity. With the accompanying loss of organization comes a tendency to lose what has been accumulated. For example, fire may release the nutrients accumulated in trees. The nutrients can then leach out of the soil to disappear into watersheds, fertilizing rivers, lakes, and ultimately the oceans. Most nutrients, however, are retained until the cycle restarts and the system reorganizes for a new phase of pioneering and consolidation.

Normally we tend to focus on the first two phases, pioneering and consolidation, when the food and fibre accumulate. Since these periods last the longest and are evident in our immediate environment, they also seem to be the only phase of any productive interest.

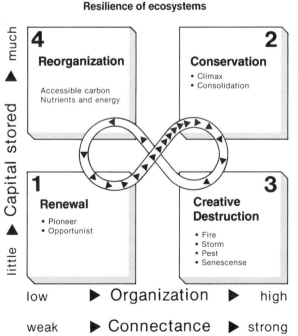

FIGURE 77. *The cycle between the four stages of ecosystem development repeats itself again and again, with the fastest growth from renewal to conservation.*

It takes a Douglas fir forest on Canada's West Coast four to five hundred years to mature. By comparison fire, insect outbreak, or wind damage happens in an instant, and though these surprises are highly visible when they occur, we tend to forget them because they are so rapid. And yet in those episodes of collapse and sharp reorganization lies the greatest opportunity for change. It is then that a rare species can suddenly take hold and stake its own territory. A new future can be generated, with new dominant species.

Any ecosystem, therefore, contains a set of unexpressed templates—patterns for alternative futures. It is during abrupt unexpected events that the rigidities are removed and the rare species, the previously unfit, can suddenly explosively co-opt the potential. This is the time for true evolutionary structural change. That is one reason why species diversity is so important—it presents ready-made options for new developments in times of crisis.

## MANAGEMENT

When people today manage ecosystems for human benefit, they try to increase the efficiency of capital accumulation to maximize production, whether of trees or crops or animals. Hence the normal abrupt changes in ecosystems are viewed as intrusions, and management considers itself to be succeeding when they are controlled and eliminated. That is why so much science and technology are directed towards control of insect pests or forest fires. But the important point is not whether we have the power to kill insects and minimize fires, but what happens when we succeed in doing so.

In every instance when we are successful in counteracting the onset of accidents, we freeze the ecosystem at some point in its phase of accumulating ecological capital. North American forests, for example, have been shaped by periodic forest fires as part of the normal rhythm of growth and change. When we try to eliminate fires we simply allow fuel to accumulate, so that when a fire inevitably occurs, the consequences can be all the more devastating. For instance, in 1988 smoke drifted as far as Toronto from forest fires that swept nearly half of Yellowstone National Park. Several thousand firefighters could not stop them, and the fires were put out only when snow and rain arrived in the fall. These fires had actually been a long time coming. For eighty-six years, from 1886 to 1972, all fires had been suppressed in the Park. As a result, vast stores of deadwood piled up, feeding the fires when they came.

Insect devastation is another example of 'surprise' when we try to freeze the ecosystem cycle. Every forty years or so an outbreak of spruce budworm used to spread through eastern Canada into Maine.

Before management, it would cause up to eighty per cent mortality of the fir trees in much of that region. We try to control a problem like that in order to achieve some social and economic objective—in New Brunswick to maintain pulp-mill employment, for instance.

There is a characteristic pattern to this kind of management. First, a policy is developed to reduce the variability of whatever is bothering people, be it insect pests or fire. Second, there is initial success. Modern industry and science, given such a simple goal, can do an extraordinary job in mobilizing the necessary technology and administration. Third, as success becomes more and more apparent, resource users become dependent on continued success. More pulp mills are built as spraying of spruce budworm seems to prevent insect outbreak. Tourism expands when fires are controlled in national parks. The result of success is a demand for more management.

But this form of management in effect freezes the system in the conservation phase. Its very success means that the ecosystem becomes more and more fragile. But that fragility is hidden by a variety of subsidies—of energy, of pesticides, of economic capital.

*Almost half of Yellowstone National Park's five million hectares were burned or damaged by fire in 1988. Naturalists believe the fires were long overdue and even necessary. (US National Parks Service Photo by Jim Peaco.)*

More pesticides must be applied to continue to prevent outbreaks of budworm. Forest-fire control must be improved to prevent catastrophic fire. This process leads to the ultimate pathology, in which the management institution becomes more rigid and the ecosystem more fragile. And the surrounding environment—society, all of us—becomes more and more dependent on this form of management. A paradox emerges. Successful resource regulation seems to lead to fundamentally unsustainable systems, but more regulation seems to be necessary.

It is partly the observation of these natural cycles and resulting pathology that has led segments of the environmental movement to oppose any kind of development in the hope of at least preventing the erosion of those properties that preserve resilience and hence contribute to sustainability. But somehow that perception seems skewed. It denies that change, in ecosystems as in society, is inevitable. No, the resolution of the paradox is to view the development of ecosystems and of the people within them in a different way, as a play. The setting for the play is the evolutionary and cultural context that allows and encourages and sustains experiments in theatre.

In this view the developmental side of sustainable development must be essentially evolutionary, providing opportunities for the next generation. And what must be sustained are the foundations of capital that can be mobilized for different purposes. In an ecosystem essential capital is made up of the soils in our prairies and our forests, the environmental health of our lakes and air, and the accumulated diversity of habitats and species. The rare species that barely survives in one habitat can potentially become a dominant organism of an ecosystem at a later time. We need, therefore, to protect our accumulated biophysical capital, preserving our capability for change and opportunity when inevitable surprises occur.

Because humankind has increasingly become a dominant force on the planet, sustainability also requires social capital that permits people to respond positively to change and ensure that surprises become not crises but opportunities. This includes the accumulated knowledge, wisdom, and skills of people in every walk of life. And culturally it pre-eminently includes social trust—trust in family, trust among the governed and of the governors.

Nations enjoying relative economic and environmental health tend to be those that have economic and social equity, and in which basic freedoms are respected. Social trust, an essential element of a sustainable society, requires some measure of equity and freedom. In many nations lack of economic equity results in exploitative use of land, both by wealthy landowners and by poor farmers forced onto marginal land. Individual initiative and co-operative groups, crucial to fostering sustainable development, are impeded by restrictions on personal freedoms. The importance of these elements of social trust is shown by Costa Rica, which leads Central America in economic equity, personal freedom, and protection of the environment.

Soils, species diversity, environmental health, human knowledge, and social trust—these are the constant foundations for sustainable development that must be protected and enhanced. We do not want to design a world that predetermines the kind of play our children will mount. We want a world in which the props and stages are available to design the alternative plays that will give them joy and opportunity.

Humanity, once only one species among many, is responsible for acid rain, for the risks of nuclear power, a reduced ozone layer, greenhouse gases, and climate change. We have been acting unconsciously as uncontrolled forces on the planet. The new direction is clearly towards viewing humanity as an integral part of a global ecosystem. Not as manager. Not even as steward. But in conscious partnership with all other life.

*Among the elements that must be protected for sustainable development are soils, species diversity, and social trust. Villagers meet in the village of Faramana, Burkina Faso, to discuss their participation in a rural-development scheme. (World Bank Photo.)*

## THE GLOBAL ECOSYSTEM

Since long before humankind became a planetary force, the planet has acted as a globally connected system. The El Niño provides a demonstration. This is an oceanic event off the coast of Peru, occurring roughly every seven years, usually around Christmas, and every twenty or so years it is particularly strong, triggering extraordinary impacts throughout the world (see pages 71-2). In 1982-3, for example, El Niño was a factor in droughts and extensive fires in Australia, a collapse of grain production in some parts of the world, extraordinary storms on the western coast of North America, and the rare appearance of certain fish stocks in northern regions. In some places the changes offered sudden new opportunities; tuna could be caught off the coast of Vancouver Island. In others, the adaptive capacity of societies came close to being overwhelmed, as in attempts to control the fires in Australia.

Such natural events seem impossibly distant from their cause half a world away. Atmospheric processes, and links between these and oceanic processes, shorten the distances. But the atmosphere has also been modulated by life: the seasonal growth in the boreal forests shows its effects in the regular annual fluctuation in carbon dioxide, and atmospheric carbon dioxide is partly regulated by marine plankton communities.

The globe has always been an interconnected system, mediated through atmospheric and ocean processes. Human activity is now affecting the environment on a global scale, unconsciously modifying those regulatory forces and becoming part of them. We can see that we are all neighbours, all affected by events half a world away.

With these insights, the lessons we have learned about evolution and ecosystems can now be focused to a new purpose, nearer to our present reality. To do this we must not only gather knowledge but must also understand the consequences if our actions are directed only to increasing economic wealth rather than to maintaining and enhancing the essential foundations of sustainability.

## GOOD NEWS STORIES

When we look in this new direction, we find some 'good news' stories in which the foundations of sustainability are being restored. They are local but they are spreading, and they furnish the hard-won experience that we can build upon for global partnership. They show how one element of surprise, crisis, can lead to opportunities for positive change towards sustainable development.

There is a lovely little alpine village, the highest settlement in

Austria, called Obergurgl. Those who live there are typical mountain people. They have lived in this valley, now four or five hours' drive from Innsbruck, for centuries. They have experienced enormous fluctuations of nature and the periodic collapse of their resource base, but they have learned to cope with change and have survived.

Immediately after the Second World War, these highly adaptive, innovative people began a new business, tourism—building hotels, ski-lifts, and all the wonderful trappings of the industry. But the glow did not last. Obergurgl became a microcosm of a global problem: population and economic growth, together with diminishing resources. These came to the surface in 1973 when sudden oil-price increases threatened tourism. At the same time the young people were getting upset with their elders, who had already built their hotels and were saying 'Don't overbuild.' The farmers were worried about too much tourism causing erosion of the mountain slopes and alpine vegetation. The traditional balance of economic and social groups, essential to a small stable community, was threatened.

*Obergurgl is an alpine ski village in Austria in which the people took responsibility for their own sustainable development. (Photo Austrian Tourist Office.)*

Then Walter Moser, a plant ecologist of enormous wisdom, at this time at the University of Innsbruck and now at the University of Alberta, perceived that what the people needed was not a solution but a pause, to give them time to step back, define their own situation, and seek not outside designs but ways to generate their own solutions, as high-mountain people had done throughout their history.

In 1971 Moser began a study of the environmental management of the Obergurgl region as a project within the UNESCO Man and the Biosphere program. With funding from the Austrian and Tyrolean governments, he brought together scientists, including a group of Canadians then working at the International Institute for Applied Systems Analysis in Vienna, and people from the village to generate alternative models of Obergurgl's future. Debate and discussions resulted, leading to a cathartic and critical town-meeting. Young and old, farmers and hoteliers, the burghermeister and representatives of the Austrian and Tyrolean governments, all debated their conflicts and asked of the models, 'What would happen if . . .?' From that inquiry came the pause needed for the identification of problems, the gradual protection of some aspects of sustainability, and the enhancement of others that had been lost, among them social trust.

But social trust was reinstated. The hotel owners set aside a fund for further research and to help villagers in ecological awareness and the evaluation of tourist reactions to the environment. They became their own researchers. The fund also provided financial support for the farmers, not as an act of charity but to help maintain the necessary diversity of social life. The interdependence of different groups in the community was recognized and strengthened.

Obergurgl supported an experiment where people themselves became the architects of their own sustainable development. This achievement was of far greater importance than the stabilizing of the local economy. The community and the individuals in it were both enriched and strengthened. This kind of local action can be the beginning of planetary change. The lesson learned there has since spread throughout the Austrian and French Alps. Scientists from elsewhere have visited Obergurgl to learn about the process, and the results have also been applied to provincial regional planning. Moreover, students of the Canadians involved have since formed a company, a team of ecologists and modellers, to help people develop their understanding of their own problems in a range of situations in Canada, the United States, Asia, and Europe.

Initiatives like this—and there are many—tend to be precipitated by crises. Perhaps the most severe crisis today is in the African Sahel—the semi-arid region stretching from Senegal to Ethiopia. Over the last few years television has shown the agony of millions of formerly self-sufficient people of this region. They have seen their land turn into wasteland, their livestock and even their children die, and they have been forced to travel to relief camps for food. What appeared on television, however, was only the last stage of a process of impoverishment that had been going on for many decades.

The Sahel was originally a productive region. Complex systems of agriculture were maintained by farmers and pastoralists who had great knowledge of their natural environment and were well able to adapt to semi-arid conditions and occasional droughts. But these systems were disrupted through war, the slave trade, and colonial occupation. More recently, poorly planned development projects ignored both the needs and capabilities of the people and the ecological limits of the land. The result has been a loss not only of soil but of the cultural foundations for sustainability: many stable communities have been destroyed, and immense knowledge of the local environment, invaluable for designing agriculture adapted to the difficult African environment, has gone to waste. Where once the people and their environment could weather occasional droughts, now this resilience has been lost.

*In Kenya continued soil erosion threatens the country's agriculture. Near Lake Baringo, the wind whips the thick dust into a storm every afternoon. (World Bank Photo.)*

The tragedy of the Sahel is that the crisis was not inevitable. The challenge is that the crisis need not continue. The knowledge exists that can turn crisis into opportunity: to change destructive practices, to build on local expertise and experience, to restore social trust, to develop ways of living that are both ecologically sound and socially equitable. There are some signs that this knowledge is being applied.

Let's look at one hopeful example from Africa. The Government of Kenya recognized in the 1960s and early 1970s that continued soil erosion was threatening the country's agriculture and asked Sweden for assistance. The Swedish International Development Authority responded by assigning a geologist, Carl-Gosta Wenner, to help the Kenyan Ministry of Agriculture find solutions. Beginning in 1974 he studied existing farming methods in Kenya and then developed strategies of soil conservation, based on traditional practices, that required only simple tools, could be done during the dry season when farmers had time available, and would not only halt erosion but usually increase yields. A network of government workers have combined with existing traditions of collective self-help and women's groups to share the expertise and labour. The program has been very

*Strategies of soil conservation based on traditional practices, simple techniques, and self-help have improved two out of five farms in Kenya. (World Bank Photo.)*

successful. Since 1974, 365,000 Kenyan farms, two out of every five, have been improved. Only moderate funding has been necessary, as the program relies as much as possible on simple techniques and self-help. It has spread quickly through education and because it provides both long-term sustainability and immediate gains in yield, and therefore in farmers' incomes. As a result, farmers have been anxious to adopt the methods.

The process by which the Kenyans are achieving success is happening in many locations throughout both the Sahel and the rainforest, and indeed worldwide. An essential first step is full understanding of both the region's ecology and the needs and abilities of the people. Progress has been most rapid when aid enables people to look past the daily struggle to survive, to use and share knowledge they already possess, and when local applications are emphasized.

There are situations, however, in which environmental degradation is so great that community or individual action alone is not enough. The challenge then is to establish links between local action and regional or global concerns. Acid raid is one such issue. Another is the spruce budworm problem in eastern-Canadian provinces. Because budworm spraying engendered community confrontation in New Brunswick, the provincial government was forced to develop a coherent regional approach to all aspects of forest management. As a result, New Brunswick now has the most robust, insightful forest policy of all Canadian provinces.

Yet another example of local problems needing regional attention is the degradation of the Great Lakes ecosystem. As the population of the region has grown over the last 150 years, the Great Lakes have become progressively degraded (Figure 78). Municipal, industrial, and agricultural pollution has been discharged; coastal wetlands have been filled in by urban development; fisheries have been depleted; dams have blocked streams; and detrimental foreign species, such as the lamprey, have entered the lakes. Since before the turn of the century there have been scattered attempts to deal with local degradation, but only in the 1950s did a public awareness begin to form that the entire Great Lakes system, and particularly the lower lakes, have become seriously degraded. Since then perhaps $10 billion have been spent in both Canada and the US on attempts to deal with this. Most have focused on single problems, such as lamprey eradication, or control of phosphates pollution.

It is becoming more evident that this approach will simply ensure further environmental degradation, since—especially in the areas most affected by human activities—the degradation can rarely be traced to a single factor. When one form of pollution is controlled, another takes its place. For example, efforts to protect the fishery by

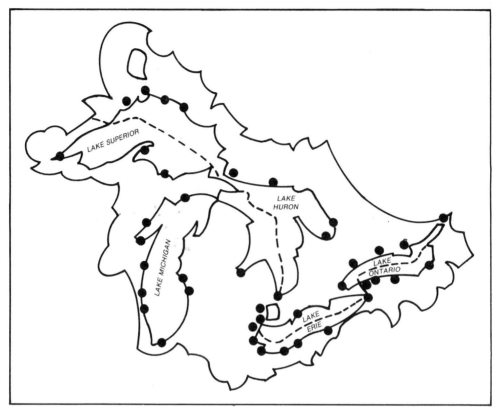

FIGURE 78. *Areas of concern in the Great Lakes Basin of North America, which has become progressively degraded over the past 150 years. While local areas of concern have been identified, Great Lakes management needs an ecosystem approach. (International Joint Commission.)*

eliminating lamprey populations are nullified by possible toxic contamination of the fish.

As a result, in 1978 the Great Lakes Research Advisory Board of the International Joint Commission (IJC) called for an ecosystem approach to Great Lakes management that would take into account the entire region and all factors that affect the lakes. Since then, progress on implementing the ecosystem approach has been slow, full of obstacles. Jurisdiction over the Great Lakes region is shared between the two national governments, several state governments and the province of Ontario, and numerous local governments. None have sufficient authority to impose a regional perspective. There also continued to be a priority, particularly in government and business, for short-term economic gain over longer-term sustainability and ecosystem integrity.

There are, however, signs of progress. Many no longer view a degraded environment as an inevitable price of economic progress but realize that economic and social vitality depends at least partly on a healthy local ecosystem. The prominence of environmental issues in both Canadian and American elections shows the public's concern. And the recent emergence of basin-wide problems—such as high water levels and toxic waste contamination—are pushing the public to view the lakes as a single interconnected system.

In 1985 the Great Lakes Water Quality Board of the IJC designated 42 severely degraded 'areas of concern' for priority attention. These sites—including both Toronto and Hamilton harbours in Canada, and the Detroit and Niagara Rivers separating the United States and Canada—are not only polluted locally but are also sources of contaminants for much of the rest of the lakes. Action will benefit both the local area and the entire basin, demonstrating the link between local and regional environmental quality, while providing a forum for citizen involvement.

Perhaps the most significant progress is the development of regional networks of people and organizations that can partly substitute for the lack of a single regional authority. New interest groups range from local groups, such as Friends of the Spit, which urges protection of the Leslie Street Spit near downtown Toronto, to larger bodies such as Great Lakes United, formed in 1982 to develop unified positions on regional issues. Together these groups help promote a regional perspective, while enlarging communication between officials of each jurisdiction. At a time when governments are under pressure to retrench and cut spending, these unofficial communication networks may become more important as an alternative to relying on government to expand its responsibilities whenever new problems appear. Through such groups and their networks, linking local issues to regional concerns becomes more practicable. The region's governments have seen the need for a regional perspective to economic and environmental policy: in 1985 the governors of the Great Lakes states and the premiers of Ontario and Quebec signed the Great Lakes Charter, which recognized the integrity of the Great Lakes Basin as a single entity. Thus, while the Great Lakes region is not yet being managed as a single ecosystem, the first steps at least have been taken.

These examples suggest some of the ongoing necessary actions that are being taken towards sustainable development. They include preservation of the foundations of sustainability, from fertile soil to healthy lakes and rivers. Another element is the enhancement of accumulated knowledge and social trust, both of which are crucial to human society's capacity for change towards ecologically sound behaviour. Developing this capacity for change may take many forms. In Obergurgl it was the evolution of means for a community to help determine its own future. In Kenya it was the adaptation of traditions of soil conservation and self-help to ensure long-term survival of the country's agriculture. In North America it is the evolution of a regional perspective of the Great Lakes environment, and of institutions that can begin to restore health to these degraded lakes.

Deterioration of the global environment is accelerating, and is accompanied by a declining standard of living in Africa, and now in Latin America. More surprises—like the discovery in 1985 of the Antarctic ozone hole—probably lie ahead. But there are some indications that society is capable of rapid, positive change. The recent pledge of the United States government to act promptly to control acid rain—after eight years of inaction—is one promising example. A second is the momentum for protection of the ozone layer. The 1987 Montreal Protocol, itself a significant achievement in limiting production of CFCs, is being surpassed by some countries that are moving towards their complete elimination. A third is the recent change in attitude of many African governments towards population control; once indifferent, many now see it as crucial for national survival.

These examples suggest that humanity is able to change its behaviour to avoid disaster. But the trend of environmental deterioration indicates that only a few years remain for such changes to occur before there is permanent damage to the earth's capacity to support life.

# CHAPTER 15

# Peace, Security, and New Forms of International Governance

## FEN OSLER HAMPSON

It was in the mid-1980s that the scientific debate about global change began to be a topic of growing interest for the general public. This was when there surfaced the suspicion that human activity was somehow behind crippling drought in the Canadian and American prairies, hurricanes in the Caribbean, and floods and famine throughout Africa and Asia. Warnings from scientists that pollution from human activity and the build-up of greenhouse gases in the atmosphere were changing the world's climate finally struck a responsive chord.

The media echoed and fuelled the concern: 'too hot to work . . . too hot to shop . . . Lobster catches died while being transported to market . . . scores of illnesses and death' (*New York Times*); 'Flood-ravaged Sudan faces threat of cholera . . . two million Sudanese . . . homeless by the floods' (Toronto *Globe and Mail*); '. . . thousands of people, leading cattle and carrying what they could . . . converging on military outpost in northern Bangladesh . . . their houses washed away' (International *Herald Tribune*); '. . . four million acres of land burned this year . . . only heavy rain or snowfall will snuff the devastating forest fires' (*New York Times*).

These quotations are symptomatic of the new threats to security that humanity faces by climatic change. They tell us what could happen: loss of life and homes; famine and pestilence, along with more refugees; intensified economic problems and conflicts; threatened national survival; and the possible collapse of international order. The social, economic, and political consequences of climatic change will be global in scope, with profound implications for every country in the world. It will require international co-operation, the rich nations helping the poor.

## ENVIRONMENTAL CHANGE AND HUMAN HISTORY

The human species has had to live with environmental change throughout history, and when it has been drastic, the human response has been to move to more hospitable regions. A volcanic eruption about AD 700 in what is Canada's Yukon devastated hundred of thousands of square kilometres, and forced some tribes of the Athapascan peoples to migrate. In Europe a thousand years later, a warm, moist summer activated the Irish potato blight of 1846. The ensuing famine and typhus killed a million people in Ireland, and with the mass exodus to the New World halved Ireland's population by the end of the century. In the American Midwest drought from 1932 to 1937 turned the plains east of the Rockies into a dustbowl. Hundreds of thousands of ruined farmers left, some to form new population centres on the west coast.

Sometimes threatened people have not had the opportunity to move. Civilizations have been destroyed by violent changes in the earth and atmosphere. The eruption of the Santorini volcano in the Aegean Sea may have caused destruction of Minoan civilization on the island of Crete. Another volcano in AD 900 destroyed existing societies in what is now El Salvador.

Environmental fluctuations have affected the economy and changed the way people earn their livelihood. In the fourteenth century a sudden cooling trend broke a warm regime that had lasted through much of the eleventh to thirteenth centuries. Some years the grain harvest failed to ripen across most of Europe, and sheep and cattle were plagued with diseases that swept the wet landscape. Most serious, the bubonic plague arrived in Europe in 1348-50 and killed off about a third of the population. The disease originated in China or Central Asia, after flooding and exceptional rains in 1332, and killed millions in Asia before sweeping Europe and signifying the end of the medieval age.

Between the 1920s and 1950s a period of North Atlantic warming boosted fish stocks off the western coast of Greenland and the coasts of Iceland, but oceanic cooling in the 1950s eliminated cod stocks around Greenland and cut Iceland's yields to less than a third of that in the 1930s. Depleted yields and overfishing fuelled the cod wars between Britain and Iceland in the 1960s and 1970s, each asserting sovereignty over North Atlantic fishing rights.

In the past we were powerless to prevent changes in the atmosphere around us. The same cannot be said today. Human activities are directly responsible for changing the composition of the Earth's atmosphere. Dramatic environmental change will cause major disruptions to the world economy, exacerbate North-South divisions, and increase the likelihood of political and military conflict, not only within states but also internationally.

Changing precipitation patterns and a rise in sea levels, coupled with rapid population growth in many regions of the world, can jeopardize global food security through shifting and diminished agricultural production, especially in marginal producing regions, hurt prospects for sustainability economic development and poverty reduction, and increase instability and the potential for international conflict. What are the specific potential global, social, economic, and political consequences of climatic change? How is environmental change likely to affect economic, social, and political relationships?

We can visualize conflicts within and among nations whose populations are multiplying while basic food and fresh-water supplies are diminishing, and further conflict over access to vital resources, like fresh water, as weather and rainfall patterns shift. Ongoing disputes over fresh-water resources, like the Indo-Pakistan or Iraq-Syria water-boundary disputes, may be aggravated. Some 214 river basins are shared by two or more countries and twelve river basins are shared by five or more. Even in the absence of climatic change, pressure on these resources from rising populations will grow (see Chapter 7).

## DESERTIFICATION AND FLOODING

Environmental degradation like desertification, deforestation, and flooding will create millions of new refugees. Whole nations may disappear with rising sea levels (see Chapter 2). A sea level rise of one metre, for example, will displace an estimated 8 million Egyptians alone. About twenty-seven per cent of Bangladesh would be threatened by a sea level rise of one metre, immediately affecting twenty-five million people.

Bangladesh's flood disaster in the late summer of 1988 was a disturbing forewarning. Two-thirds of Bangladesh was flooded by torrential monsoon rains, and the Ganges and Brahmaputra Rivers overflowed, reaching their highest levels ever. The region of Dhaka was inundated, and the floods left some 25 million people—roughly a quarter of the country's population—homeless. With water supplies polluted, an outbreak of gastric diseases and diarrhea threatened those who had escaped drowning, starvation, or snakebite. As the floodwaters retreated, many wondered if the tragedy could have been averted. Monsoon floods have worsened because trees in the Himalayan foothills of northern Bangladesh, Nepal, and Northern India are being cut for fodder and fuel. Rains surge across deforested areas, carrying topsoils into rivers, causing massive silting and flooding hundreds of miles downstream.

We might see future beach and cliff erosion due to rising sea levels in other coastal and population centres. In South America, for example, heavily populated areas like Rio de Janiero, Brazil, and Mar del Plata, Argentina, would be hit. If the sea level rises one to two metres, sea water will pollute surface and ground water and threaten deltas, coastal marshlands, and embankments. The mangrove belt along the coasts of most tropical deltas will deteriorate or even disappear. Many vulnerable areas in developing countries are dependent on agricultural production and therefore especially open to socio-economic disaster. Physical disasters exacerbate political and social upheaval caused by economic inequity and population pressure.

*Two thirds of Bangladesh was flooded by torrential monsoon rains in the summer of 1988, leaving 25 million people homeless—an indication of what could happen if the sea level rises. (Roger Lemoyne CIDA/ ACDI photo.)*

*When the Himalayan foothills of northern Bangladesh, Nepal, and northern India are deforested, rains surge across the areas, flooding hundreds of miles downstream. (World Bank Photo.)*

## REFUGEES

The economic and social disruption following climatic change indirectly contributes to political revolution. The sub-Sahara drought of the early 1970s, climaxing in 1972-3, killed an estimated 100-to-200 thousand people from the Sahel to Ethiopia and caused mass migrations southwards. Coffee crops in Ethiopia dried up, and the stresses of famine helped trigger the revolution that toppled the imperial dynasty of Haile Selassie, fuelled a civil war, and continues to rack the country.

The world has approximately twelve million trans-border refugees. Many have been uprooted by wars at home, such as the five million

*The world has about twelve million trans-border refugees, including these in a UNICEF relief camp in Uganda, who are fleeing drought and famine as well as civil unrest. (World Bank Photo.)*

Afghanis living in Pakistan and Iran. Others are also victims of drought and famine, such as an estimated 625,000 in Ethiopia who have sought to escape war and drought-stricken regions in neighbouring Sudan and Somalia, or the 167,000 in Algeria leaving the Western Sahara.

The cumulative social, economic, and political impact on 'environmental refugees' place a heavy burden on the host countries. Although a number of treaties and international agreements prohibit the return of refugees to their countries of origin, many countries, including Canada, are adopting rules, laws, and policies to restrict their admission.

Dr Jorge I. Hardoy, of the International Institute for Environment and Development in Buenos Aires, writing about the 1982-3 drought in the highlands of Bolivia and Peru, describes the problem of rural/urban migration for developing countries: 'A drought affects a city in many ways. One of the first visible impacts is the flow of peasants from the rural areas. After consuming their scarce grain reserves— even seeds for the first crop—and killing the few animals they have, peasants . . . moved to the city of Potosí [Bolivia] and many migrated to La Paz or to Argentina and Chile. Some household heads remained with the hope of saving their herds. The results of such migrations in the education and health of children is easy to understand. I visited Potosí in 1984 and jobless migrants were visible everywhere. . . .'

Because many developing countries depend heavily on subsistence and cash-crop agriculture, they will be especially vulnerable to environmental change (Figure 79). The key factors governing their ability to change are population growth and food production. From 1970 to 1982 the world's population grew at a rate of 1.8 per cent per year, but food production grew faster, by 2.3 per cent. Nevertheless regional disparities and major distribution difficulties caused famine and starvation in much of the world. If the world's population continues to grow to somewhere between 8 and 10 billion by 2025, demand for food will obviously increase.

At the same time, food production may be jeopardized by changing precipitation and climatic conditions, especially in mid-northern latitudes, and further south if America's great plains become unsuitable for producing wheat. More variable weather in semi-arid and marginal areas could also reduce crop yields. Farmers in most areas will need more irrigation water because of rising temperatures.

Stratospheric ozone depletion could damage plants—including rice, other food crops, and trees—and possibly ocean zooplankton and phytoplankton, important to the marine food chain and commercial fish stocks. New Zealand officials already report some preliminary signs of these effects.

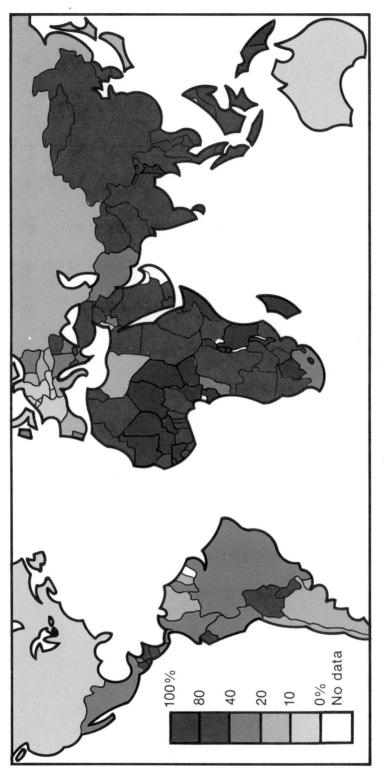

FIGURE 79. *Percentage of population dependent on agriculture, 1980. Heavily dependent countries will be especially vulnerable to global climate changes.* (Third World Atlas, *Open University Press.*)

The legend reads:

100%
80
40
20
10
0%
No data

## CHANGING GLOBAL POWER BALANCES

Global change will inevitably alter economic and political power balances. Major grain producers, such as the Soviet Union, could find their dependence on international markets increasing if changing climate threatens local food production. The Soviet Union is especially vulnerable to climatic changes, suffering major grain-production shortfalls because of weather fluctuations. In 1972, and again in 1980, it was forced to make major grain purchases from Western countries. Climatic change could bring severe drought to the Eurasian plains and severely hurt, if not eliminate, grain production in parts of the Soviet Union (though other areas may become more productive). At the same time, traditional grain exporters—like Canada and the United States, who have been able to make up their trade deficits in the Soviet Union and Eastern Europe—could find their surpluses diminishing if arable acreage shrinks with environmental change. These scarcities would increase internal instability with unpredictable, potentially grave results.

While some countries might suffer, others might build power with population and economic shifts. Imagine this scenario, given by Francis Bretherton of the US National Center for Atmospheric Research: 'August in New York City. The temperature is 95 degrees; the humidity 95 per cent. The heat wave started on July 4 and will continue through Labor Day. While warmer temperatures might boost the fish catch in Alaska and lumber harvests in the Pacific Northwest, the Great Plains could become a dust bowl; people would move north in search of food and jobs, and Canada might rival the Soviet Union as the world's most powerful nation.'

Changing climate might also increase international conflict in the Arctic and Antarctic, especially in areas where sovereignty claims remain unresolved. Atmospheric warming would melt summer pack ice, leaving much of the Arctic Ocean ice-free, opening Northeast and Northwest passages to longer shipping times. But increased viability of the Canadian Arctic in terms of transportation, mineral exploitation, and even perhaps agricultural production, might initiate challenges to Canada's sovereignty in its northern territory. Unresolved sovereignty claims in Antarctica might also be exacerbated if climatic change made this region more accessible. The military-strategic importance of both polar regions might increase with easier access. Though these scenarios are speculative, they could follow the climatic changes many scientists believe will happen.

## NATIONAL AND INTERNATIONAL RESPONSES TO ENVIRONMENTAL CHANGE

What is to be done? How should the nations of the world respond to global change? National and international policies to deal with climatic change must include a combination of 'adaptive' and 'preventive' measures. We cannot afford to underestimate the potential negative costs of warming. Enormous social costs accompanying flooding of coastal and farmland areas, for example, will inevitably widen the gap between developing and developed countries.

At the national level we will need co-ordinated action across all levels of government. Some key questions:

• What are the costs of action and inaction? Who will bear them? If the burden falls on the shoulders of the federal government, how will it respond to increasing demands when its own revenue base may be shrinking because of environmental damage to the national economy?

• How will local, regional, national, and international responses be co-ordinated?

• What are the most salient factors in the determination of the resource allocation of national governments? How can environmental issues be raised on the agenda of all levels of government: federal, provincial or state, and local?

Atmospheric change will affect every country in the world, requiring co-ordinated and concerted international as well as national action. There are sound reasons for Canada to assume a prominent role in the organization of a global-response strategy—not least of which is the environmentally sensitive Canadian economic base. The Canadian people and the government were strongly committed to the work of the World Commission on Environment and Development, which was set up as an independent body in 1983 by the United Nations, and headed by Madame Gro Harlem Brundtland, Prime Minister of Norway. Its brief was to re-examine the critical environment and development problems on the planet and to formulate realistic proposals to solve them. The Commission's 1987 report, *Our Common Future*, demands a marriage of economy and ecology, so that governments and their people can take responsibility not just for environmental damage but for the policies that cause the damage, some of which threaten our survival.

## A LAW OF THE ATMOSPHERE?

Global change presents new opportunities for international co-operation between East and West, North and South. The atmosphere is a 'commons' and must be treated that way. We may need a 'law of the atmosphere' to help govern and care for this global commons.

The history of Law of the Sea negotiations gives us some idea how challenging attempts at international regulation can be. The proposal for a new Law of the Sea was first made by Dr Arvin Pardo, the Maltese delegate to the United Nations, in 1967. He urged that the ocean sea-bed beyond the limits of national jurisdiction be declared 'the common heritage of mankind' and that the mining of the sea-floor to exploit rich manganese nodule deposits be undertaken on behalf of the international community. His concern was two-fold: first, to prevent open conflict between nations exploiting these resources; and second, to ensure that the developing countries shared in the benefits. A year later the *ad hoc* Committee on the Seabed of thirty-five nations, including Canada, considered the structure and powers of a new international organization to oversee the exploitation of the seabed resources of the international area. At the same time it was widely recognized that the enormous increase in the use of the oceans for commercial and military purposes, fishing, energy production, and scientific research had led to serious international frictions and conflict, emphasizing the inadequacy of the existing international law relating to the sea.

To address this situation the UN General Assembly convened the Third Law of the Sea Conference in 1973. By 1978 these negotiations—the largest and longest-running international negotiations ever to take place—had reached agreement on most of the contentious issues, including a 12-mile territorial sea, a 200-mile exclusive economic zone for coastal states, and provision for unimpeded transit through straits and archipelagos. However, a number of issues remained intractable, including the issue of how to establish and finance an international seabed authority.

The challenge was to resolve the views of the industrialized countries, which had the financial capacity and technological expertise, and the developing countries, which had the votes in the UN. An operating arm of the Conference, called the Enterprise, was proposed to undertake mining in international waters. For every area mined by a state or multinational entity or consortium, an equivalent area would be set aside in a land bank for the Enterprise, or for developing countries, for eventual exploitation. The 'have' countries would contribute financially towards beginning mining operations and get the Enterprise on its feet. The result would be a parallel, or mixed, system of development. However, mining interests that had already spent millions of dollars developing deep-sea mining technology were jealous of their investment and concerned about transferring their technology to the Enterprise.

Although by 1981 the Law of the Sea Treaty was ready for signature, the United States refused to sign it on the grounds that the provisions would set 'dangerous precedents' in US relations with the Third World. Nevertheless, the United States has abided by most of the Treaty's provisions, many of which had already been established as precedents in international law.

The Law of the Sea negotiations failed to establish an international seabed mining authority. However, the Treaty succeeded in codifying and legitimizing the new international regime regarding the rights of coastal states for jurisdiction in the 200-mile economic zone, and freedom of transit through international straits. And its process of development—international agreement through discussion and negotiations, rather than through warfare—is a model.

The question arises whether the Law of the Sea could be a useful model for a new law of the atmosphere. There are some differences. First, before the Law of the Sea negotiations began, there was a consensus about the need for a new regime of the oceans. In the case of the atmosphere, an international consensus had not, in 1989, been developed. Second, there was already a well-established body of international law and a partial regime in place dealing with the oceans. The issues dealt with in 1973 were simply an expansion of this regime and led to the development of additional rules, norms, and principles. In contrast, the bed of international law concerning the atmosphere is not well defined and must be created from the ground up. Third, although huge multinational companies and consortia are involved in the exploitation of ocean resources, as they are with the degradation of the atmospheric commons, there nevertheless seems to be a recognition that survival of the human race, not just material profit, is involved in degradation of the atmosphere. Some big chemical companies are taking a lead in research for alternatives, and in limiting or stopping production of life-endangering products like chlorfluorocarbons and other chemicals.

These similarities and differences offer both advantages and disadvantages in developing a law of the atmosphere. As with the Law of the Sea, arriving at an international agreement between so many different countries, and rich and poor, and with so many issues, will take all the skill negotiators can muster, and much time—perhaps more time than we have. Interim actions to slow climatic change therefore become all the more important.

## INTERIM ACTIONS

The 1988 world conference on 'The Changing Atmosphere: Implications for Global Security', which met in Toronto, was a major milestone. The immediate measures identified there included imple-

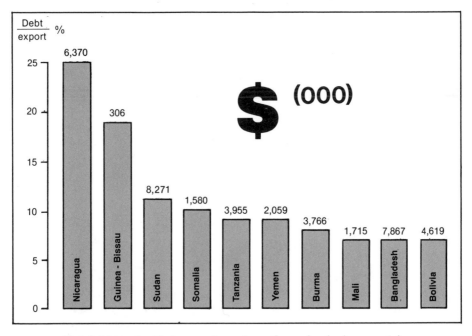

FIGURE 80. *Total debt and ratio to export earnings. Some developing countries owe so much that export earnings cannot cover their debts. (From* The New Internationalist, *November 1988.)*

menting proposals to protect the ozone layer, reduce greenhouse warming, and share the costs, all with the goal of protecting the atmosphere (see Chapter 2 for details). The development of global and national action plans was proposed.

Since the poorer nations are likely to sustain the heaviest losses in the interaction of population growth with atmospheric and ecosystem deterioration, we will have to pay special attention to their needs and transfer resources from rich to poor (Figure 80). For developing nations to forgo the use of fossil fuels in their industrial development, or the use of other fuels, such as wood for domestic purposes, will be politically sensitive, difficult, and expensive. Why should they make this sacrifice if the rest of the world is not setting an example? Likewise, there is no way to halt deforestation in order to reduce $CO_2$ increases without substantial development assistance to the Third World countries. Again, why should they deny themselves practices that northern nations have already indulged in? And why should they suffer without recompense?

Some countries have called for debt relief to developing countries so that they can afford to halt deforestation (in many cases this is

*For developing nations to forgo the use of fossil fuels or wood for domestic purposes would be difficult, expensive, and politically sensitive. Firewood is already sparse in Burkina Faso, West Africa. (World Bank Photo.)*

**Debt Outstanding**

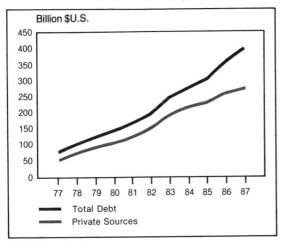

**Debt Service**

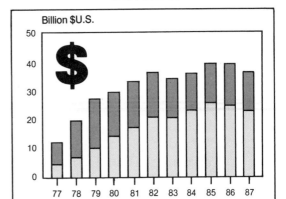

FIGURE 81. *Highly indebted countries, with a high ratio of debt service to debt outstanding, must take longer to pay off the debts. Their only recourse is to cash in on their natural resources, including vulnerable forests. Some countries have called for debt relief so that they can afford to halt deforestation. Highly indebted countries include Argentina, Bolivia, Brazil, Chile, Colombia, Costa Rica, Côte d'Ivoire, Ecuador, Jamaica, Mexico, Morocco, Nigeria, Peru, The Philippines, Uruguay, Venezuela, and Yugoslavia.* (World Debt Tables, *IBRD/World Bank Washington, 1988-9.*)

the only solution to debt costs). (Figure 81.) It is true that debt relief in the form of outright cancellation of outstanding loans, more generous terms of interest, and extension of existing loans may help reduce pressures for development. And there is a rationale for this step in that the original debts were encouraged by the rich countries looking for a place to invest their oversupply of oil money at good interest rates. However, we must also recognize that the financial problems of most developing countries are rooted in population pressures and poverty. In Brazil, for instance, ninety per cent of arable land is concentrated in the hands of two per cent of the population, who drive poor and landless peasants into jungle areas. Therefore only domestic political reforms, including changes in land ownership and tenure policies, will halt deforestation. Within those changes, poverty-stricken sharecroppers and landless peasants, as well as big landowners, will continue to burn the forests for a few years of cultivation or cattle grazing.

Righting such wrongs is ambitious but highly necessary, demanding far greater attention than is now given to them nationally and internationally. However, there are encouraging signs. The World Bank recently initiated hybrid loans, a mixture of financing for policy reforms and projects within a sector. And in 1985 the Bank—along with some rich-country donors, including Canada—set up with the government of Ghana a Program of Action to Mitigate the Social Costs of Adjustment. This program funds various social projects to help groups most hurt by the economic and public-policy reforms Ghana is undertaking.

In May 1989, 86 countries—including India and China—addressed directly the deterioration of the atmosphere, deciding at an international forum in Helsinki to eliminate by the year 2000 production of CFCs and other chemicals eating at the ozone shield. This commitment goes well beyond the Montreal Protocol of 1987, which called for CFC reductions of fifty per cent, and followed earlier commitments by Canada, the United States, and the European Community. The accord also called for industrialized countries to create a special UN fund to help developing countries industrialize without CFCs, which are used, among other things, for refrigerator coolants and plastic foam. Norway's environment minister pledged 0.1 per cent of the country's gross national product to this fund if other countries would follow suit. A group has been established to work out the details of this fund.

In the campaign against acid rain Canada, the United States, the countries of Scandinavia, the European Community, and Eastern Europe signed a protocol in November 1988 to hold the line on nitrous oxide emissions while working toward more stringent reductions. Some countries (Canada, Germany, the Netherlands, and the Scandinavian countries) have already applied higher standards.

Finally 37 countries are participating in an Intergovernmental Panel on Climate Change sponsored by the World Meteorological Organization and the UN Environment Program to follow up on the recommendations of the 1988 world conference on 'The Changing Atmosphere', and to prepare for a Second World Climate Conference, to be held in Geneva in 1990. The aim: to develop the principles and components for a convention on climate change to be tabled in 1992.

These initiatives can be strengthened by new international institutions to co-ordinate responses to climate change. For example, the Soviet Union has taken an idea proposed by Western countries—including Scandinavia, Canada, and other UN members—to convert the UN Trusteeship Council into an Environmental Trusteeship Council, where high-level officials could review and co-ordinate international responses to environmental concerns. Strengthening existing institutional machinery while creating new institutions will require financial commitments. Unfortunately, in response to domestic pressures many countries, including Canada, are cutting overseas commitments, and existing agencies like the United Nations and the UN Environment Program are having troubles matching deed with rhetoric. Nevertheless these encouraging developments demonstrate that the global community can work together. Global ecological collapse, with resulting international chaos, is not inevitable if nations address seriously the problems of global climatic change. Although hard choices and sacrifice are required, the consequences of inaction are surely worse. Only radical change in human activity and national values will save the planet's life-support system.

| MILITARY EXPENDITURE 1984 | As % of GNP | Amount (US$ mil) |
|---|---|---|
| World | 5.6 | 768,833 |
| Developed | 5.6 | 618,849 |
| Developing | 5.6 | 149,984 |
| **Rank** | | |
| 1 Iraq | 50.0 | 14,000 |
| 2 Oman | 27.7 | 2,108 |
| 3 Israel | 27.1 | 6,427 |
| 4 Saudi Arabia | 21.7 | 22,570 |
| 5 Yemen, Arab Rep. | 17.6 | 742 |
| 6 Yemen, PDR | 17.0 | 191 |
| 7 Syria | 16.6 | 2,735 |
| 8 Angola | 14.2 | 988 |
| 9 Jordan | 14.1 | 559 |
| 10 Iran | 13.3 | 10,700 |
| 11 Libya | 12.9 | 3,700 |
| 12 Nicaragua | 12.4 | 402 |
| 13 USSR | 11.5 | 225,400 |
| 14 Mongolia | 10.5 | 200 |
| 15 Chad | 10.4 | 63 |
| 16 North Korea | 10.2 | 2,270 |
| 17 Somalia | 9.6 | 168 |
| 18 Ethiopia | 9.3 | 442 |
| 19 Egypt | 8.5 | 2,917 |
| 20 Brunei | 7.9 | 304 |
| 21 UAE | 7.4 | 1,932 |
| 22 Lebanon | 7.3 | 440 |
| 23 Greece | 7.2 | 3,049 |
| 24 Cuba | 7.1 | 1,350 |
| 28 US | 6.4 | 237,052 |
| 30 UK | 5.4 | 26,525 |
| 40 Australia | 3.4 | 5,709 |
| 52 Canada | 2.3 | 7,780 |
| 54 New Zealand | 1.9 | 500 |

| HEALTH EXPENDITURE 1984 | As % of GNP | Amount (US$ mil) |
|---|---|---|
| World | 4.1 | 565,503 |
| Developed | 4.8 | 527,096 |
| Developing | 1.4 | 38,407 |
| **Rank** | | |
| 1 Sweden | 9.0 | 9,290 |
| 2 West Germany | 8.1 | 54,482 |
| 3 Ireland | 7.5 | 1,363 |
| 4 Iceland | 6.8 | 228 |
| 5 Netherlands | 6.7 | 9,092 |
| 6 France | 6.7 | 37,149 |
| 7 Canada | 6.5 | 21,499 |
| 8 Norway | 6.4 | 3,783 |
| 9 Italy | 5.9 | 23,107 |
| 10 Belgium | 5.9 | 5,066 |
| 11 Denmark | 5.8 | 3,400 |
| 12 Switzerland | 5.6 | 5,870 |
| 13 Australia | 5.5 | 9,375 |
| 14 Finland | 5.5 | 3,026 |
| 15 UK | 5.4 | 26,525 |
| 16 Czechoslovakia | 5.3 | 5,090 |
| 17 Saudi Arabia | 5.0 | 5,243 |
| 18 New Zealand | 4.9 | 1,277 |
| 19 Austria | 4.7 | 3,248 |
| 20 Japan | 4.6 | 56,874 |
| 21 Nicaragua | 4.6 | 148 |
| 22 US | 4.3 | 159,500 |
| 23 Poland | 3.9 | 6,730 |
| 24 Greece | 3.6 | 1,543 |
| 25 Israel | 3.5 | 834 |
| 27 USSR | 3.2 | 62,700 |
| 36 Iran | 1.6 | 1,286 |
| 41 Ethiopia | 1.4 | 66 |
| 52 Iraq | 0.8 | 224 |

SOURCE: Table II, *Comparative Resources, World Military and Social Expenditure SIPRI Yearbook 1988: World Armaments and Disarmament*, Oxford University Press, New York, 1988.

# The Challenge of Environmental Peacemaking

The Economic Summit meeting in Paris in July 1989 was nicknamed 'the first green summit'. Despite sustained efforts by many in Canada, environmental issues had received but a fleeting reference at the previous summit meeting of the seven western powers, held in Toronto in June 1988. A week later, in the same city, the Prime Ministers of Canada and Norway pledged personal commitment to action in front of delegates drawn from over sixty countries to the international conference on The Changing Atmosphere. The Montreal Protocol on ozone (see Chapter 2) was signed into history, and the tough reality of legislating controls on fluorohydrocarbons had begun. In Paris in 1989—during another long, hot summer—the compelling political appeal of voters' enthusiasm and media coverage led Western governments to declare war on environmental degradation. But to turn the Paris *Communiqué*, a document of seven pages and 19 clauses, into real action by some 167 countries, rich and poor, is daunting.

The interplay between rights and responsibilities, and economic, social, cultural, moral, and spiritual progress, has been the overriding theme of governance—and of war and revolution—since recorded history began. Today we are struggling with what seems like an invisible global war, whose tangible environmental consequences are blots on various continents and whose solution is environmental peacekeeping on an unprecedented international scale. In this contest, just what *is* the role of government? Indeed, which governments and alliances of government can create the necessary climate of legislation, regulation, and resource allocation? Which of our institutional mechanisms are appropriate to the multisectoral and multilateral global challenges ahead? How do we avoid entrenched local interests in reaching solutions that make global sense and command respect for action? We must ensure that our race against time in enacting solutions to problems of climate change does not become a surrogate activity for familiar ideological, territorial, jurisdictional, or bureaucratic arguments.

There is always a difficult line to draw when value judgements are made, or when values are placed upon private and public goods. In the context of sustainable development, revisiting

North-South or East-West arguments will not be as productive as will recasting all of our thinking into a longer-term integrated perspective, where economic progress and environmental issues are everybody's business. The global 'commons' are not just the air, land, and sea of our immediate personal experience but include, for example, outer space and deep ocean space, as well as the frequency spectrum for global communications technologies. In adjusting today to all the constraints, how do we get advice that is insightful, informed, objective (and unsentimental), and turn it into timely policy action—into concrete steps whose visible results may not appear for years?

The answer is leadership. The role of government in science is not just to fund research—it is to be a positive influence in the shift in thinking we all must make to save this planet. Until recently there has not been much public clamour for this shift—and therefore not much political will. This is changing. Now action can be crafted, at least at home. The first step is to ensure a shared and common vision, articulated in a clear and uncompromising way, by the most senior levels of government. The second step is to translate that vision into a set of strategies, where local actions contribute to a co-ordinated and coherent national, and ultimately international, impact. This is now the time for new alliances between scientific, business, labour, and government sectors.

A national action plan is a momentous task at the best of peaceful times. An international action plan—built through a blend of consensus, negotiation, and single-minded leadership—will require global statesmanship to ensure that the 'umbrella convention on climate change' of the summit has lasting results. But before an action plan can be drafted, policies based on knowledge, and on a mutually understood state of uncertainty, must be in place. In research laboratories, *getting the question right* is the major step forward in arriving at a potential solution.

At future gatherings of world leaders, science and technology must become an intrinsic part of the diplomatic tool-kit, to join with economics, business, law, and trade. The technologies for defence and tactics of military procurements can also be readily marshalled for environmental monitoring and resource management, as can many developments from the research laboratories of the past three decades.

We have substantial scientific knowledge, but there are still conspicuous gaps in our global data bases, monitoring and reporting mechanisms. To apply our knowledge, we need an alliance of science, industry, and government—areas that have so far been traditionally isolated or ignorant of each other. The Organization for Economic Cooperation and Development (OECD) has been trying to develop a meaningful set of environmental indicators. This is critical for those policy-makers who must advise on strategic decisions on, for example, water resources, land management, export and energy policies. Governments everywhere, at all levels, are involved in trade negotiations. I see an international forum for their participation on environmental impact in those negotiations. We can urge government policy-makers to negotiate in all these economic arenas on behalf of action *for* the environment, and demand performance that is ecologically sustainable as well as open, fair, and internationally competitive. This, however, is a long-range goal.

The Canadian National Round Table on Environment and the Economy, formed in 1989 and reporting to the Prime Minister, has adopted the goal of sustainable development. From education to setting pollution standards, this group is engaged in a very careful examination of the way we live and do business, considering not only regulatory practices but also taxes and subsidies, foreign-policy implications, demonstration projects, economic incentives and market disincentives. In the meantime the private sector is responding to the public mood. Green bonds, environmentally friendly investments (whether in stock portfolios or in the contents of grocery bags), blue boxes for recycling, issues of business journals devoted to health-care products and environmental concerns—all can be taken as signs of a change in attitude as well as a recognition of potential profits. But creative public policy is needed to address a 'disposable' mindset, given the scarcity of landfill sites, and the NIMBY (Not In My Back Yard) syndrome. There is, however, an almost naïve (but honest) belief in the public mind that energy will abound, lights will stay on, and manufacturing and service industries (thus jobs) will grow without, for example, major hydro or water-dam diversion projects, or enhanced nuclear energy. 'Demand management' can stave off the next energy crisis, but only for a few years. Alternative energy sources will then be gratefully accepted by all people who have taken energy, like water, for granted. We responded fast to the oil crisis of the early 1970s. Can we do the same now?

This is a global theme: what should the future energy policy be? Do we have the national and international will, between developed and developing countries, to tackle all at once the problems created by acid rain, carbon dioxide and nitrogen oxide emissions? What is the role of nuclear energy? The debates, and the silences, reveal much about our national character and our national will to act for the future—for to work internationally, we must begin by co-operating nationally. In Canada this is a perennial challenge.

It is human folly, not malfeasance, that has led us to the environmental and economic concerns we are grappling with at the close of the twentieth century. As we enter the next century, I hope that those young people now sitting in classrooms, who are excited by the future, will be able to say that leadership, a grasp of the issues, and the necessary actions are both evident and effective. Will love of our planet, and common sense, succeed in the early 1990s? Despite some cynical views, and deep expressions of private angst and hopelessness, we must not give up our responsibility to act with courage now. There are positive fluctuations in the present policy vacuum, and public expectations now are running high.

Thinking of all the unexpected natural catastrophes the Earth is heir to—from volcanic eruptions to earthquakes and tidal waves—I hope that people by the year 2030 will not only take care of our planet but will also take heed of those other planets we will then be exploring in our solar system. The equilibrium ecology, natural catastrophes, and climatic changes of 2030 may well demand a fresh way of thinking about the forces that govern planet Earth. In the meantime, the revolution in thinking and action we require *now* demands a profound change in economics and structural adjustments, and in the models, assumptions, and values by which we live. We are faced with nothing less than a litmus-test of our very human stuff.

GERALDINE KENNEY-WALLACE

# CHAPTER 16

# Summing It Up

## J. STAN ROWE

Modern physical and biological sciences are revealing the vitality of the planet and the relatedness of all its parts, people included. At the same time the unexpected harmful environmental effects of technologies that these same sciences have created are forcing into our consciousness a new appreciation of the worth of the natural world in all its beauty.

Through the use of science/technology we have enriched ourselves and materially improved our welfare, but with high costs to the Earth-source. Those costs can no longer be deferred. Each new inventory shows Earth's renewable resources diminishing in quantity and quality. Pollutants in food, air, and water seem to be unavoidable. The health of our 'environment'—vaguely defined and dimly sensed—is being ruined. No clear escape-route from these threatening circumstances is in sight.

I believe that answers will come from a more profound leading vision as to what 'environment' really is, guided by deeper insight and greater affection for the surrounding world. This in turn will lead to informed actions, for when we get our values and perceptions straight, everything else falls into place. Busy-ness, unguided by enlightened priorities among those things we consider most important, can be fatal.

In youth we are all fitted with cultural belief-spectacles that synthesize, focus, and colour a particular world view that seems both self-evident and 'right' for those born to it. When, as today, signals from the real world clash with the expectations of the world view fabricated by our particular culture, then the time is right for a 'new look', a change to better glasses, keener insight, improved theories about where we are, who we are, and what we should be doing.

Already we can glimpse in outline some parts of the emerging new look, for now we know the planet as a miraculous floating cell that supports people as conscious parts of its fertile skin. If our sense of time could be collapsed from antiquity to today, we would see in an instant our relationships from the beginning to air, water, soil, and myriad other organisms. All are parts of the marvellous Earth-system, companions on an evolutionary journey whose end is unknown.

In a similar way, if we could see, all at one time, our dependent relationships to the different parts of the Earth, we would also recognize the ecological truth that we are inseparable from our surroundings. How foolish attempts to dominate, control, manage, and reshape the Earth-environment in the immediate interests of one single species would then seem! We would understand that waging war on our environment for short-term gain is suicidal.

Today's appropriate world view questions the widespread Western masculine belief that the leading purposes of thought and action are to master and to manage. It offers instead a goal of compliance, co-operation, and understanding to which the foregoing chapters all bear witness. Let us review what they say.

Complex problems that afflict the world—the subjects of this book—are accurately described as 'messes'. Such are global changes in air chemistry, in climate, in vegetation, population, economic development, militarization, and run-away technology. Most messes are related, they seep into and get mixed up with one another, and the word reminds us that in the face of such confused complexity, simple analyses and quick-fix solutions are futile. In short, the human condition is more complicated than we know, and perhaps more intricate than we will ever fully grasp.

People-induced messes are superimposed on a planet whose evolution and future course are far from fully understood. Thus the range of concerns, the hopes and fears, the irritation and patient indulgence, the optimism and pessimism, expressed or implied by the contributors to this book. Yet all are caring people, and all recognize that humanity must soon take a new path, different from the one followed in the past. Digby McLaren's Preface identifies the global problem as a lack of balance between people and the world environment, brought on by rampant population growth and profligate use of non-renewable energy. Though ignorant of how to run the planet, we nevertheless are forcing our will on it. William Fyfe takes up the same theme in the opening chapter, tracing the energetic history of the globe and the recent assaults on its diversity by the one species that has the power of reflection. He poses the question in the minds of all contributors: Can we find it within ourselves to harmonize with the processes of the Earth?

The thorough review of climatic change and its industrial causes by Gordon McKay and Henry Hengeveld in Chapter 2 brings into sharp focus the need for public policies and immediate action if the atmosphere overhead is to be protected. John Matthews reinforces the message in Chapter 4. Comparing past climatic change, as indicated by temperature-related glacial-age strata in northern Canada, with that of today, he concludes that human activities are changing climate more rapidly than at any other time in the last 100,000 years. Richard Peltier in Chapter 3 attributes unexpected past responses in climate and sea levels to small variations in the tilt of the earth's axis or in the radiation received from the sun—another indication that incremental changes in the atmosphere, brought on by naïve industrial pursuits, may cause sudden unexpected flips in our Earth environment.

An examination of the polar regions, the oceans, fresh water, forests, and grasslands reinforces the fact that industrial society has become a potent geological force affecting every sector of the world. In Chapter 5 Fred Roots describes the 'ends of the Earth', the Antarctic and Arctic, and the particular sensitivity of these polar regions to lower-latitude industry. David Schindler and Suzanne Bayley in Chapter 7 examine the relationships between climate and fresh water, which is the life-blood of the land and the usual limiting factor to its productivity. With or without climatic warming and drying, Canada's northland will be the target of grandiose water-diversion plans because of its many lakes and rivers that flow away from the population concentrations of the south. The authors rightly warn that such schemes will mean the impoverishment of the North. Water 'management' is perhaps the classic example of how we solve perceived social and environmental problems: not by engineering our behaviour but by engineering the resource.

In Chapter 6 Lawrence Mysak and Charles Lin identify ocean-circulation patterns as an important transport mechanism for heat, affecting the chemistry of the water (the uptake and release of carbon dioxide) and the world fisheries. The last section of their chapter examines predictive models, a subject meriting close scrutiny because on this one peg hang all the dire warnings of catastrophic climatic change. How reliable are the General Circulation Models (GCMs) that forecast global temperature increases of 1.5 to 4.5 degrees centigrade when some time in the next century the carbon dioxide content of the air will double to 600 parts per million?

The first unsettling fact is that all GCMs are incomplete; they do not take full account of the effects on climate of the seas, of cloudiness, and of vegetation. Why, then, should anyone believe them? The answer, and the second disturbing fact, is that society demands prediction from science 'or what good is it?' Having accepted the honour and adulation merited by discoveries that contribute to wealth and health, scientists find it difficult to admit that prophecy, particularly with respect to complex systems, is not their long suit. Modelling that began as a tool for explanation has been pressed into service as a crystal ball from whose blurry depths probable futures are read. Let buyers beware of exaggerated claims, even as they accept the attention-getting utility of exaggeration!

In Chapter 8 Jag Maini tells the story of the world's forests, whose depletion began with the use of fire and with the clearing of land for agriculture. As integral parts of the global ecological system, forests not only react to the climatic regime but also affect it. The simplest and most direct way to lessen carbon dioxide in the atmosphere is to plant trees world-wide—a program that would, however, be costly in terms of capital and land. Slowing the cutting of mature forests—anathema to the forest industry—is an alternative that ought to be considered, though it rarely is. Grasslands, like forests, also respond to changes in temperature and precipitation patterns, and in Chapter 9 John Stewart and Holm Tiessen warn of the probable decline of agriculture in semi-arid climates wherever the warming trend is felt.

The destruction of forests and the deterioration of agriculture in grassland regions are obviously happening today—not waiting on climatic change. Both rich and poor countries abuse their forests, just as wealthy farmers and poverty-stricken peasants alike abuse their land. The downward trend in land-use preceded the build-up of greenhouse gases, and it is likely to continue regardless of climatic warming, which will mainly serve to accelerate it. Humanity's numbers, needs, and wants are the problem—a topic dealt with in the final section.

Part IV shifts attention from the ecosphere and its ecological systems to one important component: the species of featherless biped, which none too soon shows signs of entertaining the notion that things other than itself have also been 'made in God's image'. These chapters examine people—their motives, beliefs, and actions that cause conflict both within the human community, and between it and the planetary environment. The driving forces are value-based and strongly biased towards material well-being. The goal in developing countries is simple survival; in industrial countries it is the accumulation of wealth; and on the international stage, need clashes with greed.

Peter Timmerman, in his enlightening Chapter 10 on ethics, identifies a wrong way of thinking at the root of the people-in-the-world problem. How are we to change? Are we to return to values we have somehow lost, or find new values, or simply live up to those we now only profess? He favours a combination of the first two. The worthwhile values we lost are those that encouraged our ancestors to tune their behaviour to the patterns of the universe. The new values we must find are the counterbalance to rights: namely, responsibilities, duties, and obligations—not just towards other members of our race but, more important, towards the natural world, negotiating a global contract with Nature.

In Chapter 11 Susan McDaniel treats the population mess. What interests her primarily is not the question of a world-people balance, but the demographic process: what makes it tick, why people act as they do. Why do birth rates rise and fall? What social and economic forces explain large families and small? We don't know, she says, nor do we know the carrying capacity of the Earth; but meanwhile, in both industrial and developing countries, people—especially young people —are needed for economic growth. This message runs counter to the theme that many of the stresses our planet is suffering are ultimately the result of over-growth and over-consumption by the human species. And surely the degradation of the world, visible wherever we look, is proof that the world's 5.2 billion people are at least twice too many—until genetic engineering cuts our average height down to 30 centimetres or less, thus reducing our needs!

It seems to me that it is time to reject humanistic cant about 'people, our most valuable resource' and start shouting that human over-population is a central problem in both industrial and Third World countries, intensified by the amount of energy used per capita. We can agree with McDaniel that the ways to control and reduce human numbers are legion, and that education is mandatory. In this regard, and setting a primary goal for education, the special role of women as conceivers and nurturers of infants lends urgency to the task of ensuring their full and equal participation—politically and economically—in all societies.

In Chapter 12 Donna Smyth examines technology, whose good side is its encouragement of a global consciousness, and whose bad side is its reinforcement of the status quo, strengthening the power of the wealthy who have easiest access to it. Believing that the use of technology is based on non-technical choices, she expresses cautious optimism that the right choices can be made, and so technology can provide the means to the creation of a better world. Kenney-Wallace makes similar arguments for science.

It is a mistake to assume that science and technology do not contain an inner drive and direction, even though both are instrumental means. The fact that they are devised to serve humanity first indicates that by their very nature, in an unthinking way, they do great disservice to the non-human world. Until the hidden agenda of traditional science and technology is exposed, they are part of the problem, not the solution. Whenever, as now, we get bogged down, and someone shouts 'More research will get you out!', we should pause and ask what the contribution of the last round of research was to creating the bog? The ozone holes from which science will rescue us were created thanks to the diligent research of an earlier generation of innocent scientists.

Taking the superficial view, that science and technology are neutral means, the key to judging their goodness or badness is whether or not they are in harmony with the ends they are supposed to accomplish. What, in fact, is 'the better world' that technology will foster? Is it a world of more people engaged in more material economic growth—efficient, competitive, profitable, and underwritten by the public through government fiscal policy, as Roy Aitken argues in 'Industry's Response' following Chapter 13—or is it a healthy and beautiful ecosphere populated by manageable rather than escalating numbers of *Homo sapiens*?

In Chapter 13 Robert Goodland and Herman Daly partly answer this question, not by championing ecological goals but by criticizing traditional theory directed to pure, sweet, unlimited economic growth. They point out that the economic system is first of all a belief system historically rooted in an era when both human population and industrial development were comparatively small. The central preoccupation in 1776, when Adam Smith published *The Wealth of Nations*, as now, was growth—an expanding scale of resource use. But the early economic system operated in a seemingly infinite environment from which resources could be freely taken, and into which wastes could freely be expelled. In contrast, today's global economic system is out of proportion compared to its environmental host. It has become abundantly clear that the economic system must be fitted to the capacity of the ecosphere to sustain it.

C.S. Holling and Stephen Bocking stress in Chapter 14 our involvement in change and evolution, and the need to maintain flexibility and adaptability for whatever surprises may befall us. The best preparation for the future is to preserve the resiliency of complex organized systems: both the natural ones from which we draw food, clothing, and shelter, and the diverse cultural ones of different nations and ethnic groups.

International co-operation is imperative, and Fen Hampson calls in Chapter 15 for a global response to meet global problems. National selfishness in a small world is a time-bomb—potentially the most explosive issue threatening the whole human enterprise. We need to make hard choices, national sacrifices, radical changes in human activities, and—coming full turn to the beginning of the book—changes in human values. The important challenges Hampson names are the same ones posed by the World Commission on Environment and Development—the Brundtland Commission—whose members solicited opinions and facts world-wide from 1984 to 1986 and undertook to define a global agenda for change based on a common endeavour and new norms of behaviour. In some ways the present book, stressing the dimension of global change, is a sequel to the Commission's report, *Our Common Future*. A critical analysis of the report provides a useful supplement to the ideas of the authors just reviewed.

## SUSTAINABLE DEVELOPMENT

Sustainable development, said the Brundtland Commission, is 'development that meets the needs of the present without compromising the ability of future generations to meet their own needs'. By casting the definition in the form of a humanitarian rather than an ecological goal, by not inquiring *what kind of economic growth* may be sustainable and *what kind* may not, by expressing only general sentiments about such fundamental problems as technology, population growth, social organization, and energy use—the Commission elevated Sustainable Development to the position of unquestioned virtue recently vacated by motherhood. Conversion to the faith, when it means all things to all people, is easy. Nevertheless Sustainable Development is a useful rallying call. The most helpful response would be to clarify its creed and examine its intent, feasibility, and implications. Let us therefore ask, in a constructive sense, what it rationally can mean.

The focus of 'sustainability' is usually taken to be economic return. Agriculture can only be sustainable, some say, when its profitability is guaranteed. But to take an extreme view, what if intensive agriculture—the human enterprise of burning down forests, ploughing up land, pouring in energy and chemicals to produce high-yield crops—finally renders the global environment unlivable? The result, of course, would be disaster. *Ecological* sustainability should be the goal.

The economy operates as a sub-system within the ecosphere, from which it draws energy and materials to provide for basic needs of food, clothing, and shelter, while also attempting to satisfy the limitless wants of an entrepreneurial species. Technology is simply the means—a combination of beliefs, ideas, knowledge, instruments, and inventions—by which the economic system draws sustenance from, and expels wastes into, the surrounding ecological system on which it completely depends.

Humanity's economy is comparable to the fetus in a mother, and technology to the placenta and umbilical cord. Through technology (as through placenta and umbilical cord), materials and energy that have been organized into usable forms by the surrounding system are taken in and appropriated for the maintenance and growth of the inside system; the unused and degraded remains are expelled into the maternal system as 'wastes'. But experience and ecology tell us that there are no 'free lunches'. The larger the feast, the larger the wastes that the outer system must absorb, detoxify, and reconstitute.

Notice the close analogy between the economic system in Mother Earth and the fetus in mother animal. Just as a size limit exists for the fetus that a 65-kilogram woman can support without undermining her health, so there must be a limit to the size of the human economy that the healthy ecosphere can support. In both instances the subsystem can threaten the health of the larger enveloping system in two ways:

(i) by appropriating materials and energy faster than the parent can renew them; and/or

(ii) by poisoning the host system—expelling wastes in quantities and at rates that exceed the detoxifying capacity of the parent.

All the current environmental problems of the world come to this: that in every country, in every region, people armed with increasingly powerful technology are appropriating more and more from surrounding ecosystems to feed not only themselves but also huge ever-growing industries that excrete garbage and toxins. Population growth and industrial growth are producing massive garbojunk that is raising global temperatures, destroying the atmospheric ozone, causing acid rain, and polluting the oceans, the air, streams, lakes, soils, food, and organisms.

The eating-up and the spewing-out are mostly incremental, in little bites here and there and everywhere, making assessment of the magnitude of the danger difficult, but all together adding up to major catastrophes. Note that the problems are *quantitative* and *material*, involving greater and greater appropriation of 'raw materials' that degenerate into pollution problems. In this context, nothing good can be said for the military industry, which yearly wastes and pollutes the world at the expense of one trillion dollars of destroyed wealth that could otherwise be preserved, or at least used judiciously.

Greater efficiency in the use of materials, more recycling, clean technology, and better waste disposal—all are good as far as they go. But they lead to complacency by acting as safety valves for an outworn system, bleeding off the steam needed to power fundamental changes.

Sustainability means putting a cap on growth of resource use, but it can accommodate development of other sorts. Were *quality* rather than quantity the economic system's goal, unrestrained and exuberant 'growth' could go on forever, providing more services, more cultural activities, more learning, more musical harmony—and solutions to injustice and inequity.

Without clear priorities, humanity will destroy the world, and inadvertently itself. The idea that the two pursuits—economic development that looks after the welfare of people, their needs and wants, and environmental conservation that attends to the welfare of the world—can go along merrily hand in hand as equals is false. The *first* priority has to be the welfare of surrounding ecosystems, for without their healthy functioning no economic system can last. The world religions all condemn selfishness in individuals. We must now go one step further and recognize that species selfishness, hitherto condoned, is not only stupid, it is criminal.

We should consider that the ecosphere and its sectoral ecosystems can get along fine without humans—but we cannot survive a moment without *them*. The ecosphere made us, but we do not have a clue about how to reconstitute it once we irreversibly destroy or poison its parts. In size, age, and evolutionary creativity, it far surpasses the human race in importance.

The first step in clarifying the meaning of sustainable development is establishing as a goal the health of the ecosphere. Economic considerations must be *secondary*. The economic system must reinvent itself as a tool in the service of human and ecospheric health, discarding its present role as an exploitative means to human wealth.

Nevertheless, the most pressing question facing us is what size of population, using what kinds and quantities of energy and materials, can safely be accommodated to the Earth's healthy functioning? In imagining and preparing for the future, we must feel our way forward, seeking goals, directions, and guidelines rather than open-and-shut solutions, for we are in a new situation, glimpsing ecological insights that are less than fifty years old. We are trying to escape thousands of years of homocentric navel-gazing, of species-selfishness, of science/technology activities that have been devised to further the search for omnipotence. Their bitter fruits are evident today in what we call 'resource depletion', 'pollution', and 'deteriorating environmental quality'.

## NEW GOALS FOR SCIENCE

Conflicting ideas about saving the environment and ourselves are vying for attention. Whole sections of newspapers call on us to save the planet by reducing human pressures on it—while other sections contain hand-wringing business reports about slow economic growth. On the one hand we agree that we ought to act more altruistically and co-operatively towards the poor and the powerless, towards the developing countries and the environment. On the other hand, individualism and privatization are considered supreme civic virtues. The competitive route seems far more popular in the world of real politics, despite much evidence that co-operation is the only viable and progressive road in the future.

Some trace the cause of this disparity to the education system, but it only reflects the norms of our society. Others blame the scientific framework provided by Darwinism: the survival of the fittest. But Darwin did not invent the competitive model—it was popular long before his birth. Perhaps its roots are in the very way we perceive reality.

The objectivity of dispassionate science is a dogma widely accepted by the public, and even endorsed by some scientists—when they take time away from the emotional defence of their pet theories and hypotheses. But the separation of viewer and viewed, of subject and object, of knower and known—which in earlier times was considered the essence of the scientific method—does not stand up to close analysis. The so-called objectivity of method thus derived turns out to be subjective after all, and a deterrent to sympathetic knowing in the full sense that comprehension implies.

Immersed in the world ecosystem, we have not grasped the meaning of our true environment. We have fragmented our surroundings, and constructed fields of knowledge, disciplines, educational systems, departments—an entire culture of arts and sciences—around the fragments. But revelations from outer space of the environmental whole, interpreted by ecological understanding, are challenging age-old ideas of human pre-eminence and purpose that have brought the world to the brink.

The unity is the ecosphere—literally the home-sphere, the global 'being' whose inseparable physical/biological parts have evolved together for 4.6 billion years. Life is a miraculous product of the ecosphere, one of its important elements, not a phenomenon isolated in a hostile world. *Homo sapiens*, a single species among 20 million fellow creatures, is an active component, relying on the whole. We have a choice. Are we to be a knowing, co-operative part of the whole, or a selfish scourge? Will we be gardeners of the world or its cancer?

The wisdom we seek rides with a harnessed troika: (i) the *way we understand* and learn (the mode-of-knowing); (ii) the *things we believe* to be real and worth knowing (the content of knowledge), and (iii) the *values* and importance of both (ethics). The three are tied closely together. For example, the dominant Western mode-of-knowing—objective and scientific—is accompanied by an appropriate ethic, whose values are individuality, entrepreneurship, and self-satisfaction—the 'me separate from, and against, the world' perspective. Ethics are tuned not only to the mode-of-knowing of traditional science but also to ideas of what is real and centrally important (worth knowing), for to be ethical requires a sense of purpose. Here in the West, we have nominated ourselves as the 'real and centrally important' fact.

In their three-way relationship, important knowledge, the best way to get it, and the values attached to it in thought and action are mutually reinforcing. If *people* are the central reality of the universe, then the sole focus of ethical concerns will be *their* rights and values. If we accept these premises (and most of us unfortunately do), then scientific objectivism is the correct mode-of-knowing, for no one can dispute the exploitative power over nature that it delivers.

But if things other than humans are of surpassing importance, as today's events lead many to suspect, then a new purpose—Salvation of this Planet—is revealed, bringing into question the old ethic and the old mode-of-knowing. If any one point of the triangle is successfully challenged and changed, then the other two will readjust.

If we adopt Salvation of this Planet as our goal, then we reject power and control as our guiding light. The shift from an anthropocentric and egocentric focus to a Whole Earth or ecocentric focus will not devalue people, but will enhance their prospects for a better life. The intrinsic value of people will still be recognized, but in the context of the more important ecosphere. Sympathetic science can show the way.

For science—which was first an exploration of God's design revealed in nature, then a means for human mastery and control over nature—has inadvertently opened a new chapter in the book of knowledge whose dazzling insights can change fundamental ideas about the planet/people relationship, about values, and about the way we know the world. Science may indeed recapture its earlier promise as the instrument of human self-enlightenment by contributing to self-awareness, helping the race to feel its way towards a definition of itself as part of nature—and as an agent of change—that will enable us better to protect this one and only planetary home.

# For Further Reading

## 1. DYNAMICS OF PLANET EARTH

Clark, W.C., and R.E. Munn, eds., *Sustainable Development of the Biosphere*, Cambridge University Press, 1986.

Grove, Jean M., *The Little Ice Age*, Methuen Press, London, 1988.

Lovelock, J.E., *The Ages of Gaia: A Biography of Our Living Earth*, Oxford University Press, 1988.

Lovelock, J.E., *Gaia: A New Look at Life on Earth*, Oxford University Press, 1987.

McLaren, Digby J., and Brian J. Skinner, ed., *Resources and World Development*, Dahlem Workshop Reports, John Wiley, West Sussex, England, 1987.

Press, F., and R. Siever, *Earth*, 4th edition, 1986.

*Scientific American*, 'Managing Planet Earth', Special Issue, September 1989.

Skinner, B.J., *Earth Resources*, Prentice-Hall, 1976.

Smith, D.G., ed., *The Cambridge Encyclopedia of Earth Sciences*, Cambridge University Press, 1982.

World Commission on Environment and Development, *Our Common Future*, Oxford University Press, 1987.

Worldwatch Institute, *State of the World 1989*, W.W. Norton & Co., 1989.

## 2. THE CHANGING ATMOSPHERE

Bates, David V., *A Citizen's Guide To Air Pollution*, McGill-Queen's University Press, Montreal, 1972.

Grant, L.D., 'Health Effects Issues Associated with Regional and Global Air Pollution', *Proceedings of the World Conference on the Changing Atmosphere*, Toronto, June 1988.

'The Greenhouse Effects: How it Can Change Our Lives', *EPA Journal*, 15 (1), 1989, pp. 2-50.

Kellogg, W.W., and R. Schware, *Climate Change and Society: Consequences of Increasing Atmospheric Carbon Dioxide*, Westview Press, Boulder, Colorado, 1981.

*Proceedings, World Conference on the Changing Atmosphere: Implications for Global Security*, World Meteorological Organization, 1989.

Schneider, T., S.D. Lee, G.J.P. Walters, and L.D. Grant, eds, *Studies in Environmental Sciences 35. Atmospheric Ozone Research and its Policy Implication*, Elsevier Scientific Publishing Company, Amsterdam, 1989.

*Societal Responses to Regional Climatic Change: Forecasting by Analogy*, Westview Press, Boulder, Colorado, 1988.

'Symposium on the Health Effects of Acid Aerosols', *Environmental Health Perspectives*, 79:1, 1989, pp. 1-205.

## 4. APPROACHING TODAY

Atkinson, T.C., K.R. Briffa, and G.R. Coope, 'Seasonal Temperatures in Britain During the Past 22,000 Years Reconstructed Using Beetle Remains', *Nature*, Vol. 325, 1987, pp. 587-92.

Fulton, R.J., J.A. Heginbottom, and S. Funder, eds, *Quaternary Geology of Canada and Greenland*, Geological Survey of Canada, Geology of Canada No. 1 (in press).

*GEOS*, Vol. 18, no. 3, Summer 1989, Energy, Mines and Resources, Ottawa.

Hopkin, D.M., J.V. Matthews Jr., C.E. Schwezer, and S.B. Jouns, eds, *Paleoecology of Beringia*, Academic Press, 1982.

Kutzbach, J., and H.E. Wright, 'Simulation Models of Climate at 18,000 Years BP: Results for North America/North Atlantic/European Sector and Comparison with the Geologic Record of North America', *Quaternary Science Review* 4, 1985, pp. 147-87.

Sutcliffe, Antony J., *On the Track of Ice Age Mammals*, Harvard University Press, Cambridge, Mass., 1985.

Wright, H.E. Jr., ed., *Late Quaternary Environments of the United States Vol. 2, The Holocene*. University of Minnesota Press.

## 7. FRESH WATERS IN CYCLE

Bocking, Richard C., in *Canadian Aquatic Resources*, ed., M.C. Healey, and R.R. Wallace, *Canadian Bulletin of Fisheries and Aquatic Sciences*, 215, Fisheries and Oceans, Ottawa.

Mathews and Fung, 'Methane Emissions from Natural Wetlands', in *Global Biogeochemical Cycles*, Vol. 1, no. 1, 1987, pp. 61-87.

Schindler, D.W., 'Effects of Acid Rain on Freshwater Ecosystems', *Science* 239, 18 January 1988, pp. 149-59.

Schlesinger and Regier, *Journal of Great Lakes Research*, Vol. 13, 1987. pp. 340-52.

## 9. GRASSLANDS INTO DESERTS?

Bourliere, F., *Tropical Savannahs*, Elsevier Scientific Publishing Company, Amsterdam, 1983.

El Swaify, L., ed., *Soil Erosion Conservation*, Soil Conservation Society of America, 1985.

Fairbairn, Garry Lawrence, *Will the Bounty End? The Uncertain Future of Canada's Food Supply*, Agricultural Institute of Canada, Western Producer Prairie Books, Saskatoon, 1984.

Golley, Frank B., and Ernesto Medina, *Tropical Ecological Studies*, Springer-Verlag, New York, 1975.

## 10. GROUNDS FOR CONCERN:<br>ENVIRONMENTAL ETHICS IN THE FACE OF GLOBAL CHANGE

Callicot, J. Baird, *In Defense of the Land Ethic: Essays in Environmental Philosophy*, State University of New York Press, 1989.

Evernden, Neil, *The Natural Alien: Humankind and Environment*, University of Toronto Press, 1985.

Lopez, Barry, *Crossing Open Ground*, Vintage Books, 1989.

MacIntrye, Alaisdair, *After Virtue*, University of Notre Dame Press, 1981.

Passmore, John A., *Man's Responsibility for Nature: Ecological Problems and Western Traditions*, Charles Scribner's Sons, 1974.

Rolston, Holmes, *Environmental Ethics*, Temple University Press, 1988.

Van DeVeer, Donald, and Christine Pierce, eds, *People, Penguins, and Plastic Trees*, Wadsworth Publishing Co., 1986.

White, Lynn Jr., essay in David and Eileen Spring, eds, *Ecology and Religion in History*, Harper and Row Publishers, Inc., 1974.

Williams, Bernard, *Ethics and the Limits of Philosophy*, Fontana Press/Collins, 1985.

## 11. PEOPLE PRESSURE

Boulding, Kenneth, 'The Meaning of the Twenty-First Century: Reexamining the Great Transition', in Michael Marien, and Lane Jennings, eds, *What I Have Learned: Thinking About the Future Then and Now*. Greenwood Press, New York, 1987.

Brown, Lester O., *Our Demographically Divided World*. Worldwatch Institute, Washington, D.C., 1986.

Demeny, Paul, 'Population Change: Global Trends and Implications', in Digby J. McLaren and Brian J. Skinner, eds, *Resources and World Development*, Dahlem Workshop Reports, John Wiley, West Sussex, England, 1987.

Repetto, Robert, 'Population, Resource Pressures and Poverty', in Robert Repetto, ed., *The Global Possible: Resources, Development and the New Century*, Yale University Press, New Haven, 1985.

Solas, Rafael M., 'Six Billion and Counting', in Frank Feather and Rashmi Mayur, eds, *Optimistic Outlooks: Latest Views on the Global Future by a Galaxy of International Experts*, Global Futures Network, Toronto, 1982.

## 12. FROM TECHNOLOGICAL FIX TO APPROPRIATE TECHNOLOGY

Barney, G.O., *The Global 2000 Report to the President*, Government Printing Office, Washington, D.C., 1980.

Boulding, Kenneth, *Ecodynamics*, Sage Publications, London, 1984.

Brooks, David, John Robinson, and Ralph Torrie, *2025: Soft Energy Futures for Canada. A Summary*, Friends of the Earth, Ottawa, 1983.

Capra, Fritjof, 'Criteria of Systems Thinking' in *Futures*, 17, 1985, pp. 475-8.

Capra, Fritjof, *The Turning Point*, Bantam, Toronto, 1984.

Castells, Manuel, 'High Technology, World Development, and Structural Transformations: The Trends and the Debate' in *Alternatives*, XI, 1986, pp. 297-343.

George, Susan, *A Fate Worse Than Debt*, Penguin, London, 1988.

Hardin, G., and J. Bodin, eds, *Managing the Commons*, W.H. Freeman, San Francisco, 1977.

Harris, Nigel, *The End of the Third World Newly Industrializing Countries and the Decline of an Ideology*, Penguin, London, 1988.

Hoffman, R.B., *Overview of the Socio-Economic Framework*, Structural Analysis Division, Statistics Canada, Ottawa, 1986.

Howard, Robert, *Brave New Workplace*, Elizabeth Sifton Books, Viking Press, New York, 1985.

Menzies, Heather, *Women and the Chip*, The Institute for Research on Public Policy, Montreal, 1981.

Sale, Kirkpatrick, *Human Scale*, Pedigree Books, New York, 1980.

Schumacher, E.F., *Small Is Beautiful*, Abacus, Sphere Books, London, 1973.

Science Council of Canada Report, *Planning Now for an Information Society*, Minister of Supply and Services, Ottawa, 1982.

Shiva, Vandana, *Staying Alive: Women, Ecology and Development*, Zed Books, London, 1988.

Sivard, Ruth Leger, *World Military and Social Expenditures, 1986*, World Priorities, Washington, DC, 1986.

## 13. THE MISSING TOOLS

Daly, H.E., 'Sustainable Development: From Conceptual Theory Towards Operational Principles', *Population and Development Review*, 1989.

Daly, H.E., and J. Cobb, *For the Common Good: Redirecting the Economy Towards Community, the Environment, and a Sustainable Future*, Beacon Press, Boston, 1989.

Goodland, R., 'The Environment Implications of Major Projects in Third World Development' (9-34) in Chester, P., ed., *Major Projects and the Environment*, Major Projects Association, Oxford, 1989.

## 15. PEACE, SECURITY, AND NEW FORMS OF NATIONAL GOVERNANCE

Brown, Lester R., *State of the World: A World Watch Institute Report on Progress Toward a Sustainable Society*, Norton, 1989.

Lamb, H.H., *Climate, History and the Modern World*, Methuen, 1982.

McNeill, William H., *Plagues and Peoples*, Doubleday, 1976.

# Contributors

ROY AITKEN is executive vice-president of Inco Limited. He is a leader in mining associations and trade and business groups, both national and international. He was vice-chairman of the National Task Force on Environment and Economy.

SUZANNE E. BAYLEY, an associate professor in the Department of Botany, University of Alberta, has studied wetland ecosystems in Florida, Louisiana, Ontario, and Alberta.

STEPHEN BOCKING, a doctoral student at the University of Toronto, with degrees in zoology and in the history of science, is studying the application of ecological research to issues of environmental quality.

IAN BURTON, director of the International Federation of Institutes for Advanced Study (IFIAS) and chairman of its Steering Committee for the International Human Dimensions of Global Change Program, was appointed science policy adviser for Environment Canada in July 1989. He is co-author of *The Environment as Hazard*, published by the Oxford University Press.

JOSEF CIHLAR is senior research scientist at the Canada Centre for Remote Sensing, where he is head of the Applications Development Section. His research interests have centred on the use of space observations for land monitoring, with emphasis on vegetation and soils. He has been principal investigator in major national and international research programs.

HERMAN DALY is currently senior environmental economist in the World Bank's Environment Department. He is on leave from Louisiana State University, where he is Alumni Professor of Economics. He has written about steady-state economics, and is co-author with John Cobb of *For the Common Good*, published by Beacon Press in 1989.

ARTHUR DYKE has for fifteen years been a glacial geologist with the Geological Survey of Canada. He maps ice-age geology in the Arctic.

CLAIRE FRANKLIN is chief of the Environmental and Occupational Toxicology Division of Health and Welfare Canada. Formerly director of the Thunder Bay School of Medical Technology, Dr Franklin is on several international committees dealing with pesticides and air and water pollution.

URSULA FRANKLIN is professor emeritus at the University of Toronto, where she taught in the Faculty of Engineering. She is also a recognized scholar and spokesperson on the social concerns of technology, on the conserver society, and on international peace. She was chairman in 1977 of the Science Council of Canada committee on Canada as a Conserver Society.

WILLIAM S. FYFE, dean of the Faculty of Science of the University of Western Ontario in London, and professor of geology, is a Fellow of the Royal Societies of London and of New Zealand as well as of Canada. As one of Canada's delegates at the International Council of Scientific Unions meeting, which instituted the International Geosphere-Biosphere Program (Global Change) in 1986, he has been a leader of the project.

R.F.A. GOODLAND is chief of the America Environment Division, Latin America and Technical Department, of the World Bank in Washington, DC, and has been a leader in the move to make the Bank responsive to the economic and environmental requirements of developing countries.

FEN OSLER HAMPSON is associate professor of international affairs at the Norman Paterson School of International Affairs, Carleton University, and an associate in the Canadian Institute for International Peace and Security, both in Ottawa. He has written widely on foreign affairs and is the author most recently of *Unguided Missiles: How America Buys Its Weapons*, published by Norton in 1989.

HENRY HENGEVELD is adviser on matters relating to carbon dioxide at the Canadian Climate Centre of Environment Canada's Atmospheric Environment Service. He was part of the secretariat organizing the 1988 World Conference on 'The Changing Atmosphere: Implications for Global Security', held in Toronto.

C.S. HOLLING holds an endowed chair in ecological sciences at the University of Florida. Formerly he was professor and director of the Institute of Resource Ecology, University of British Columbia, and director of the International Institute for Applied Systems Analysis (IIASA), Vienna.

SUSAN HOLTZ is a Halifax-based consultant in energy, resource, and environmental policy issues, and senior researcher for the Ecology Action Centre, a non-governmental environment group in Halifax. She is a member of the National Task Force on Environment and Economy and of the Nova Scotia and National Round Tables on Environment and Economy.

GERALDINE A. KENNEY-WALLACE is Chairman of the Science Council of Canada, which advises the federal government on research policy and strategy. She is also a member of the National Advisory Board on Science and Technology and of the National Round Table on Environment and Economy. An international authority on lasers and optoelectronics, she has been professor of chemistry and of physics at the University of Toronto since 1980. She will become president of McMaster University in July 1990.

NATHAN KEYFITZ is head of the Population Program at the International Institute of Applied Systems Analysis in Austria and a consultant to the government of Indonesia. He is also professor emeritus at Harvard and Ohio State Universities.

ROY KOERNER, a glaciologist now with the Geological Survey of Canada, worked in Antarctica with the British Antarctic Survey and the US program. In the northern polar regions he was with the Arctic Institute of North America, and travelled by dog sled from Alaska to Svalbard via the North Pole with the British Trans Arctic Expedition.

CHARLES A. LIN is associate professor in the Climate Research Group, Department of Meteorology, McGill University. Dr Lin was previously assistant professor in the University of Toronto's physics department.

J.S. MAINI is assistant deputy minister in the Canadian Forestry Service. He was at the International Institute for Applied Systems Analysis in Vienna from 1983 to 1985 and before that was director general, Policy, Corporate Planning Group, Environment Canada. He has served in many capacities in national and international agencies, including as adviser to the InterAction Council of former world leaders.

JOHN V. MATTHEWS JR, a research scientist in the Geological Survey of Canada, specializes in the reconstruction of past terrestrial environments in order to document past fluctuations of climate and rates of climatic change. His current research focuses on Arctic areas as they were just before the start of the ice ages.

BEVERLY McBRIDE is a writer based in Ottawa. Over the last seven years she has worked in various environmental organizations in Canada and Europe and is currently at the Canadian Environmental Network.

SUSAN A. McDANIEL, a demographer, is professor of sociology at the University of Alberta. Formerly she taught at the University of Waterloo and in the Faculty of Medicine of the University of Alberta. Her areas of interest include women's health and medical issues, and her most recent book is *Canada's Aging Population*, published in 1986.

GORDON A. McKAY, now a consultant, was acting director general of the Climate Centre, part of the Atmospheric Environmental Service in Downsview, Ontario. Before that he was director of Climatological Applications at AES. Internationally respected as a meteorologist, he was the co-ordinator of the June 1988 conference on 'The Changing Atmosphere: Implications for Global Security' in Toronto.

DIGBY J. McLAREN is the president of the Royal Society of Canada, the national academy of arts and sciences, which has assumed a large role in co-ordinating major projects involving natural and social scientists and humanists from universities, government, and private research organizations. Formerly director general of the Geological Survey of Canada, Dr McLaren has also been acclaimed in international science, and is a Fellow of the Royal Society of London and a Foreign Associate of the US National Academy of Sciences. He proposed, convened, and co-ordinated two Dahlem Conferences in Berlin in 1986 on Resources and World Development, and edited the final volume, published in 1987.

CONSTANCE MUNGALL has been a writer, editor, and broadcaster of scientific information for the general public. Author of four books related to citizen action, she was editor of *GEOS*, a popular journal on the earth sciences published by Energy, Mines and Resources Canada.

LAWRENCE A. MYSAK is director of the Climate Research Group, Department of Meteorology, McGill University. A major initiative of the group is to develop a global ocean circulation model. He is also NSERC senior industrial professor of climatology.

W. RICHARD PELTIER, professor of physics at the University of Toronto, is an international authority on motions of the earth's mantle, including mantle convections, and on the dynamics of the atmosphere. He is Chairman of the Canadian National Committee for the International Geosphere-Biosphere Program and of the science committee of the Canadian Global Change Program.

FRED ROOTS, a widely experienced arctic and antarctic scientist, is science adviser at Environment Canada, a federal government department he helped to form. Previously, as senior adviser in Environment and Northern Programs at the Department of Energy, Mines and Resources, he planned and directed the Polar Continental Shelf Project. He was a participant in the first wide-ranging studies of the Canadian Arctic islands.

J. STAN ROWE began his career as a research scientist with the Canadian Forestry Service, working in Winnipeg and Ottawa. In 1967 he joined the faculty of the University of Saskatchewan as an ecologist and he has been professor emeritus since 1985. His interests in biology, landscape ecology, and ethics are indicated by various publications on forest geography, the ecosystem concept, natural area preservation, and morality in science.

DAVID W. SCHINDLER, an environmental scientist in Winnipeg at the Freshwater Institute, Canada Department of Fisheries and Oceans, when he wrote his chapter, is now teaching at the University of Alberta. From 1968 to 1986 he was head of the Experimental Lakes Project, where he studied eutrophication, acid rain, and radioactive elements. He has headed the International Joint Commission's Expert Committee on Ecology and the Atmosphere and the Biosphere, and the American Society of Limnology and Oceanography.

DONNA SMYTH is associate professor in the Department of English, Acadia University. A professional writer, her work includes the 1985 series for CBC Ideas, 'Finding Out: The Rise of Citizen Science', a documentary novel *Subversive Elements*, published by the Women's Press, Toronto, in 1986, and a 40-page booklet for the Association of Canadian Studies, *Canada and the Environment, a Citizen's Map* (1988), which is distributed by the Secretary of State.

JOHN W.B. STEWART, director of the Saskatchewan Institute of Pedology and head of the Department of Soil Science at the University of Saskatchewan, has recently organized the Agricultural Institute of Canada conference, 'In Search of Soil Conservation Strategies in Canada'. He is chairman of the Canada Committee on Land Resource Services and a member of the International Society of Soil Science and the Scientific Committee on Problems of the Environment. For the past ten years he has been co-ordinator of a Canadian International Development Agency project at two universities in Brazil.

HOLM TIESSEN is a research scientist and adjunct professor in the Saskatchewan Institute of Pedology and has extensive experience in the tropics. He has worked for the German Agency for Technical Cooperation at the Crop Research Institute in Ghana and has current projects in Ghana and Brazil.

PETER TIMMERMAN is a research associate at the Institute for Environmental Studies, University of Toronto, and IFIAS visiting scholar in the Secretariat for the Human Dimensions of Global Change Program. He has written extensively on environmental ethics, the state of environment reporting, and the impacts of climatic change.

# Index

aboriginal relationships with nature, 213
acid deposition, 78
acid fog, 23
acid rain, 59, 165, 166-7
  causes, 23, 49, 154
  effects, 23
  reduction of, 316
acid shock, 60, 130
aerial photography, 263
Africa, 5, 11
  Sahel, 197, 294-5, 306
  Savannahs, 189, 197-202
  see also Ethiopia, Kenya, Nigeria, Zimbabwe
After Virtue, 212
agriculture, 54, 126, 127-8, 201, 295, 325
Amazon rainforest, 13, 185, 187, 282
animal rights, 217
Antarctic, 112, 119, 120, 121, 126, 309, 324
  Ice Sheet, 86, 110, 111, 126
Arctic, 17, 112, 116-33, 324
  haze, 49, 60, 112, 129
Arctic Ocean, 126
artificial insemination, 255
astronomical factors re climate, 92, 107, 116
atmosphere, 19, 42, 46-79, 265
Australia, 292

Bacon, Francis, 267
Baltic Sea, 90
Bangladesh, 5, 65, 303
Barents Sea, 118, 131
Bering Land Bridge, 104
Bering Sea, 135
Bering Shelf, 130
Betsiboka River, Madagascar, 13
Bhopal, India, 248
biomass, 40, 60-1, 189
biotechnology, 254-60
boreal forest, 107, 175, 292
Brazil, 185-7
  Caatinga, 189, 202-4
  conservation projects, 280
  distribution of wealth, 315
  food, 5
  forests, destruction of, 54, 282
  grasslands, 189
  see also Amazon rainforest
Bretherton, Francis, 309
British Clean Air Act, 47
Browder, J.O., 282

Brundtland, Gro Harlem, 310, 316
Brundtland Commission, 184, 208, 210, 260, 262-3, 310, 327
Buddhism, 219
Burke, Edmund, 216
Byrd Glacier, 17

Canada, 115
  accords, 23, 316
  economics, 181-2, 309
  forests, 13, 173, 181-2
  lakes, 149, 152-6, 297
  Northern, 86-8, 96-7, 100, 118
  Prairies (Great Plains), 11, 161-4, 189, 192-7, 205
  waste, 21
  wetlands, 161
Canada Centre for Remote Sensing (CCRS), 254, 265-6
Canadian Atmospheric Environment Service (AES), 68, 146
Canadian National Round Table on Environment and the Economy, 320
Capra, Fritjof, 261
carbon cycle, 13, 143, 169
carbon dioxide, 31-2, 42, 49, 54, 69, 145, 147-8, 163, 170-1
  as a greenhouse gas, 47, 88, 108, 144, 286
  as a natural process, 169, 292
  in atmosphere, 42, 107, 114, 122, 145, 148, 177, 325
  in ice, 112-14
  in ocean, 17, 143, 147-8, 324
carbon 14 dating, 88, 99
carbon monoxide, 59
carbon reservoir, 143, 161, 169
Carson, Rachel, 44, 247
'Changing Atmosphere, Implications for Global Security, The', world conference on, Toronto 1988, 68-9, 115, 223, 312-13, 316, 318
Chernobyl, USSR, 229
Chesapeake Bay, 59
children, 5, 21, 77, 231
China, 3, 227, 302, 316
chlorofluorocarbons (CFCs), 47
  control of, 300, 312, 316

effects of, 19, 49
use of, 54, 70
see also greenhouse gases
Christianity, 214, 219
citizen scientists, 262, 267-8
climate:
  astronomical factors re, 92, 107, 116
  system, 134
Club of Rome, 23
communications, 248-50
Conable, Barber, 279
conduction, 32
conservation, 207, 261
  policy, 280
continental drift, 36, 82-4, 134
continental plates, 38
contraception, 242, 243
convection, 32, 36-8, 40, 81-2, 84
cores, ice, 101, 112-14
  sediments in, 91, 92
crop diversification, 200, 204, 205
Cuyahoga River, Ohio, 271

Darwin, Charles, 331
DDT, 247
deforestation, 13, 54, 201
  effects of, 170, 197, 279, 303
  natural causes for, 156
  policies against, 315
  rate of, 43, 168, 185-6, 270
demography, 227, 231, 239-41, 326
Department of National Health, Canada, 77
desertification, 11, 188-92, 202, 303-5
  in Africa, 197, 295
  in South America, 204, 304
disease spread, 79
distribution of wealth, 234, 252, 262, 277, 313-15
drought, 62-4, 166, 189, 200, 203
Dubois, René, 262

Earth, 31, 80-4, 96, 285
  magnetic field, 32, 36
  mantle, 81-2, 86
  orbit, 51, 92, 93-4, 107
  seasonal precession, 94
  surface temperature, 44
  tilt, 51, 94, 116
Earth Observing System (Eos), 266
ecology, deep, 216, 217

economics, 239-41, 252, 269-84, 325
agricultural, 205-6, 309
discount rates, 274-8
ecological, 279-80, 320, 327, 328
forestry, 172-3, 180, 313-15
GNP, 270, 273, 278
incentives, 284
of developing countries, 231-3, 259, 313-15
steady state, 274-8
Economic Summit, Paris, July 1989, 318
ecosystems, 286-91, 297-300
global, 291, 292
Great Lakes, 297-300
management of, 288-91
northern, 150
polar, 120-2
tropical, 185
education, 241
Egypt, 3, 5, 21, 65
Eletrobras, Brazil, 280
energy, 7, 237, 261
conservation, 261
hydroelectrical, 156
non-renewable, 323
radioactive, 31
solar, 15, 31, 118-19
environmentalism, 214-19
Environmental Trusteeship Council, UN, 316
erosion, 43, 197, 202, 293
prevention of, 168, 191
ethics, 129, 211-24, 325, 331
Ethiopia, 201, 306, 307
Europe, 302
eutrophication, 154, 164
evolution, 179, 285-300, 327
Experimental Lakes Area (ELA), 152
Exxon Valdez oil spill, 283

family planning, 235, 242-3, 254
Ferguson, Howard, 68
fire, 199, 287, 288
fish migration, 138, 158
fishing, 130-1, 137, 149-50, 156, 297, 302
cod, 137, 302
herring, 137-8
salmon, 138
food, 5, 78, 188, 202, 256-7, 258, 307
forestry, 126-7
forests, 13, 168-87, 325
and climate, 175
boreal, 107, 175
degradation of, 64
ecosystems, 286-7
extent of, 174

fires, insects, diseases, 179
industries, 172-3, 180
reproductive capacity, 179
temperate, 175
tropical, 175-6, 185
see also deforestation
fossil fuels, xiii
effects of, 54, 76, 157, 170
use of, xv, 76, 261, 313
fossils, 98
bones, 84, 102
insect, 101
plant, 101, 105-6
Fraser River, British Columbia, 138-9
fresh water, 149-67, 324
resources, 303
Friends of the Earth, Canada, 261
Friends of the Spit, Toronto, 299

Gaia, 41-2, 43-5, 216
garbage, 21, 271, 273, 329
genetic diversity, 27, 43, 171, 205
genetic manipulation, 255, 257
genetic record, 29
Ghana, 315-16
glacial refugia, 102, 104
glaciation, 84-6, 91-2, 101, 110-11
glacio-marine relicts, 154
global commons, 49, 274, 310-11
global consciousness, 239, 245, 331
global contract, 218, 220-4
global temperature, 47, 51, 56, 62, 107
control of, 185
fluctuations in, 123, 125, 175
rise in, 43, 123
global warming, 73, 148
Gobi Desert, 11
grasslands, 188-208, 325
deterioration, 191
fire, 191
types, 191, 193, 197
Great Lakes, 297-9, 300
Charter, 299
ecosystem, 297-300
Research Advisory Board (IJC), 298
United, 299
Water Quality Board (IJC), 299
green revolution, 205
greenhouse gases, 47, 53, 58, 88, 108, 123, 144, 163, 286
$CO_2$, 47, 88, 108, 163, 286
CFC, 47

methane, 47, 108, 163, 286
nitrous oxide, 47
greenhouse warming, xv, 17, 54, 107, 141, 152, 157
Experimental Lakes Area, 152-6
in polar regions, 90
Greenland, 106, 302
Ice Sheet, 86, 112, 125, 126
Gross National Product (GNP), 270, 273, 278

Haile Selassie, 306
hamburgers, 271, 282
Hardoy, Jorge I., 307
Harvard School of National Health, 77
herbicides, 27, 71, 195, 256
Himalayas, 40
Hinduism, 213
hot spots, 36
Hudson's Bay Company, 100, 152
Human Dimensions of Global Change, 67-8, 221-4
humankind, xiv, 44-5, 56, 84, 95, 109, 130, 291, 323
health, 53, 58, 59, 66, 73-9, 235
history, 302
Husain, Shahid, 280
Huxley, Aldous, 209
Hydrogen, 42
Hydrogen sulphide, 42

ice cores, 101, 112-14
ice sheets, 91-2
see also Antarctic Ice Sheet, Greenland Ice Sheet, Laurentian Ice Sheet
in vitro fertilization, 254
India, 3, 5, 227, 229, 303, 316
Indonesia, 3, 51, 100
industrialization, 209, 215, 286, 250-2
industry, 283-4
Information Age, 250
Inter Pares, Ottawa, 280-1
InterAction Council of former heads of state, 184
intercropping, 199
international co-operation, 327
development, 233, 313
governance, 301-21
peace, 301-21
International Council of Scientific Unions, xiv, 122, 222
International Federation of Institutes for Advanced Study, 222

International Geosphere-Biosphere Program, xiv, 68, 210, 221, 222
International Institute for Applied Systems Analysis (IIASA), 294
International Institute for Environment and Development, 307
International Joint Commission (IJC), 157, 183
Great Lakes Research Advisory Board, 298
Great Lakes Water Quality Board, 299
International Social Science Council, 222
International Union for the Conservation of Nature and Natural Resources (IUCN), 207
isostatic rebound, 86-8

Japan, 229
Judaeo-Christian tradition, 213

Kenya, 3, 296-7, 300
Klein, Ruth, 254
Koppen, Vladimir, 92

Lake chemistry, 153
Lake Superior, 59
see also Great Lakes
LANDSAT, research satellite, 254, 263, 266
Laurentide Ice Sheet, 92, 101, 104, 110
Law of the Sea, 184, 311-12
Lena River, USSR, 126
Liberation theology, 219
Little Ice Age, 51, 100, 177
Lopez, Barry, 218
Lovelock, James, 41-2, 43-5, 216

McGill University Climate Research Group, Canada, 146
MacIntyre, Alaisdair, 212
Mackenzie River, Canada, 126, 141, 149, 159-60
malnutrition, 5
Marine Observation Satellite (MOS-1), Japan, 265
mercury poisoning, 144
methane, 42, 47, 49, 108, 163, 286
Milankovitch, Milutin, 92-3
military:
industry, 221, 317, 329
technology, 131, 245
models, 81, 84, 88, 91, 94

atmosphere–ocean, 119, 147-8, 197
carbon cycle, ocean, 145
climate, 54, 122, 123, 177
convective circulation, 82
General Circulation, 324
sunlight model, 237
Mohammedanism, 213
Montreal Protocol on Atmospheric Ozone Depletion, 1987, 69, 183, 221, 300, 316, 318
Moser, Walter, 294
Mt St Helens, 38
mountain-building, 40
mythology, 218

Naess, Arne, 216
Nature, traditional ways of relating to, 213
navigation, 157
Arctic, 130, 135, 309
Nigeria, 3, 7, 248
NIMBY syndrome, 129
Niña, La, 72
Niño, El, 51, 67, 71-2, 135, 152, 185, 203, 292
Niño/Southern Oscillation (ENSO), El, 71-2
nitric acid: see acid rain
nitrogen, 31, 59, 194
nitrous oxides, 47, 49, 54
North America, 78-9, 102, 104, 106
Great Plains, 161-4, 189, 192-7, 205
Norway, 129, 141, 316
nuclear power, 261
nuclear waste, 247
nutrient enrichment: see eutrophication

Obergurgl, Austria, 293-4, 300
oceans, 134-48, 324
chemistry, 35, 141
ridges, 32, 82
temperature, 135
ocean-floor spreading, 35
see also continental drift
Organization for Economic Cooperation and Development (OECD), 320
Origin of Species, 331
Our Common Future, 208, 260, 262-3, 310, 328
oxygen, 19, 42, 44, 169
depletion of, 154-6
oxygen isotopes, 92, 112, 169
ozone, stratospheric, 19, 57-8, 73, 307, 316
depletion of, 271
ozone surface, 54, 59, 76-7

Pacific Ring of Fire, 38
Pangea, 83
Pardo, Arvin, 311
Peace River, Alberta, 160
People and Chips, 250
permafrost, 61, 96, 101, 102, 106, 156
personal initiatives, 70-1
Peru, 71-2
pesticides, 27, 71, 129, 247, 248, 259, 289-90
Philosophy of science, 214, 267, 326, 331, 332
Phosphates, 25
photosynthesis, 13, 42, 121, 156, 169, 286
polar axis, 118
policy, 207, 310
agricultural, 206
conservation, 280
forest, 182-3, 185, 199, 280
global, 319
pollution, 247-8, 297, 329
air, 23, 47-9, 59-60, 76-8
Arctic, 128
effect on forests, 272
ocean, 143-4
reversal, 271
water, 21, 25, 128, 271, 297
polychlorinated biphenyls (PCBs), 60, 229
population, 225-43, 326
control of growth of, 231, 233-6, 242, 300
demands of, 42, 323
distribution of, 7
growth of, xiv, 3, 202, 209, 225
of developed countries, 225, 229, 254
of developing countries, 202, 225, 230, 235, 239, 255
public policy, 68, 324
Public Policy Recommendations, 68-71

RADARSAT, 266
radioactive waste: see nuclear waste
rainfall, 189, 191, 199-200, 202, 204-5
recycling, 21, 278
reforestation, 171-2, 177, 180
refugees, 225, 233, 303, 306-7
religious traditions and nature, 213
remote sensing, 263-6
resources:
non-renewable, 269, 323
renewable, 278
river diversions, 158-60
GRAND project, 158
NAWAPA project, 159

romanticism, 215
Rowe, Christopher, 250
Royal Society of Canada, xiii, 222
Royal Society of London, 43

Sahel, 197, 294-5, 306
St Lawrence River, Canada, 144, 149
salinization, 188
satellite mapping, 32, 263
satellite observation, 7, 9, 30, 250, 254, 263-6
Saudi Arabia, 3
Scandinavia, 316
Science, philosophy of, 214, 267, 326, 331, 332
sea level, 78, 86-91, 125-6, 134
  global-warming effects on, 17, 65, 111
  natural changes in, 141
  rise in, 86, 90, 101
  thermal expansion, 125, 148
Silent Spring, 44, 247
Smith, Adam, 269, 270, 327
social capital, 290
Social Costs of Rain Forest Destruction: The Hamburger Debate, The, 282
social organization, 229, 236
soil, 11, 13
  deterioration, 192, 194-5, 197, 202, 206
  erosion, 43, 168, 191
Soil at Risk, 258-9
solar energy, 15, 31, 118-19
Sparrow, Senator Herbert, 258-9
species:
  diversity, 149, 171, 193, 199
  extinction, 272, 288
  migration, 177
spruce budworm, 288-9, 297
subduction, 38
sulphur, 165
  compounds, 38
  dioxide, 23, 166
sulphuric acid, 166
see also acid rain
summerfallow, 194
sustainability, 275, 292, 295, 299, 329
sustainable development, xiv, 328-30
  evolutionary development of, 290
  planning, 9, 70, 278

policies for, 207-8, 320
  understanding problems of, 210, 217, 277
Swedish International Development Authority, 296
Système Probatoire d'Observation de la Terre (SPOT), France, 265

Tambora, Indonesia, 51, 100
Taylor, Charles, 214
technology, 244-68, 326-7
  agricultural, 192, 195, 229
  communications, 248-50
  forest, 172-3, 180
  genetic, 179, 254-50
  information, 254, 262
  military, 131, 245
temperature:
  of fresh water, 156
  of ocean, 135
tephra, 104
Test-Tube Woman: What Future for Motherhood?, 254
thermal processes, 15
Three Mile Island, USA, 229
Tokyo International Symposium, Japan 1988, 222
tourism, 289, 293
Tournachon, Gaspard Félix, 263
tree-ring record, 99
tropical forest, 171, 175-6, 185, 271, 282
2025: Soft Energy Futures for Canada, 261

ultraviolet radiation, 46, 57-8, 73
  skin cancer, 73-4
UNESCO, 222
UNESCO Man and the Biosphere program, 294
Union of Soviet Socialist Republics (USSR), 158, 309
United Nations:
  General Assembly, 311
  Non-Government Organizations at, 262
United Nations Environment Program (UNEP), 122, 185, 207, 316
United Nations Population Fund, 5

United Nations Trusteeship Council, 316
United Nations University, 222
United States, 110, 269, 312
  see also North America
US National Aeronautics and Space Administration (NASA), 44, 263, 266
US National Center for Atmospheric Research, 309
US National Oceanic and Atmospheric Administration (NOAA), 146
urbanization, 56-7, 66, 187

values, 211-12, 316, 318, 321, 322, 325, 331, 332
Villach, Austria, 65, 122, 123
volcanic ash, 38-9
volcanism, 32, 36, 38, 39, 51, 107, 302
Vostok ice core, 112

waste, 21
water, 25, 78, 149-67, 297
watersheds, 168
Wealth of Nations, The, 269, 327
Wegener, Alfred, 83
Wenner, Carl-Gosta, 269
wetlands, 161-4, 297
White, Lynn Jr., 214
women, status of, 3, 233-4, 238, 259
work, 241
World Bank, 234, 270, 279-80, 281, 315
World Climate Program, 68
World Commission on Environment and Development: see Brundtland Commission
World Conservation Strategy, 207-8
World Meteorological Organization, 122
World Resources Institute, 59
World Wildlife Fund, 207

Yellowstone National Park, USA, 38, 288

Zimbabwe, 242-3
Zimbabwe National Family Planning Council (ZNFPC), 242-3